The Moon
& the Western
Imagination

The Moon & the

THE UNIVERSITY OF ARIZONA PRESS   TUCSON

Western

Imagination

SCOTT L. MONTGOMERY

To Kyle and Cameron,
without whom the heavens would be empty

The University of Arizona Press
© 1999 The Arizona Board of Regents

⊗ This book is printed on acid-free, archival-quality paper.
Manufactured in the United States of America

04   03   02   01   6   5   4   3   2

Library of Congress Cataloging-in-Publication Data

Montgomery, Scott L.
The moon and the western imagination / Scott L. Montgomery.
p. cm.
Includes bibliographical references (p.   ) and index.
ISBN 0-8165-1711-8 (cloth : alk. paper)
ISBN 0-8165-1989-7 (paper : alk. paper)
1. Moon.   2. Moon—In art.   3. Moon—In literature.
4. Moon—Maps.   I. Title.
QB581.M6475   1999
523.3—dc21   99-6090
                        CIP

British Library Cataloguing-in-Publication Data
A catalogue record for this book is available from the British Library.

Thou shalt draw down the Moon from heaven . . .
For whilst thou seest the lunar disc display
Such rocks and ocean depths unfathomable,
What powers prevent thy sight of worlds celestial
From tracing all their semblance to this earth?

Jeremiah Horrox, "The Transit of Venus," 1640

Aye, pardon us, O moon,
Round, bright upon the darkening!
Pardon us our little journeys endlessly repeated!

William Carlos Williams, "A la lune," 1914

Contents

Illustrations

Preface

*F*or all ages and all cultures, the sight of the full Moon rising, ablaze in the dark of the night sky, has been a captivating vision. A sense of enthrallment when confronted with this display, whether tinged with thrill, omen, or even fear, may be among the more universal of human responses to the physical universe. What this state of emotion might mean to the individual, however, as well as what s/he might actually perceive in the glowing orb itself, will inevitably vary from one people to the next, just as it has differed for the natives of Western society from one historical era to the next. The contemporary educated German or British adult sees something quite different when s/he looks at the lunar surface than did the Renaissance Florentine or twelfth-century Parisian, not to mention the aristocratic Roman of the imperial age or the pre-Socratic Greek philosopher. As a single phenomenon, the Moon seems to have been a world with a thousand faces.

Such was my own thinking when I started the present study. Now at the end of it, I realize the need for a different metaphor. Perhaps "a thousand expressions" might be better; there have not been so many faces after all, not at least in the history of the West. Strong threads of continuity bind several traditions of representation and imagination through the centuries. One of the oldest of these is the tradition we carry with us today, which we have inherited from the opening decades of modern observation, dating from the first use of the telescope. It is common these days to view history as a succession of discontinuities, across which major shifts in mind or outlook have taken place. Such fragmentation has its appeal no doubt, and perhaps its rewards as well. But there is much evidence against it, not the least of which may be found in any close look at a science such as astronomy, with its ancient storehouse of phenomena, belief, emotion, and association.

This book is an attempt to trace one item in that storehouse through nearly three millennia of conception. It is thus a work of cultural history, concerned with patterns of meaning applied to the Moon, as told through the traditions of imagery that came to envelope it and live upon its surface. Several of these traditions gained a special type of renewal in the seventeenth century when the Moon was first mapped and its features named. This marked a profound moment in lunar history, and it is one with which I have been especially concerned for its richness of cultural significance. I have also spent no small effort pursuing the lunar body as it has appeared in literature, philosophy, and above all, art. All of these realms were inseparable from the broader field of science as it treated the Moon. Indeed, it was art far more than science per se that made of the lunar orb a planet in the modern sense, a world in motion yet fixed with a geography of names, places, and thus earthly projections.

To gaze upward at the Moon, therefore, is to look deeply into a well of history. It is to peep through a keyhole leading to the heavens in the human imagination, where still larger orbits of continuity and change are to be found. But it is also an opportunity to perceive the impossibility of strict boundaries around an area of knowledge (call it "science"), which gleams magnificently with light reflected from so many other realms.

Such perceptions, of course, are by no means novel. They are, in fact, older than might be commonly assumed. This book follows distinctly in the footsteps set down more than five decades ago by Marjorie Hope Nicolson, whose elegant insights into the varied connections between scientific and literary forms of the imagination remain among the best examples of intellectual history in this area.

The writing of this book, though very much a personal affair, benefited greatly from the kind help of several friends and colleagues. David Topper read much of the manuscript and, in addition to providing me with the benefit of his expertise on art and science, saved me the embarrassment of certain errors and omissions. Albert Van Helden encouraged the project from its inception and made many helpful comments and suggestions along the way. Stanley Gedzelman generously shared his work on meteorological phenomena in art. My deepest gratitude goes to Ewen Whitaker, for granting me entrance to his "lunar museum," for filling many gaps in source material and in my understanding of lunar history, for his generous review of the entire manuscript, and for the kind hospitality he and his wife, Beryl, extended during my brief stay in Tucson. Finally, I thank my editors at the University of Arizona Press, Amy Chapman Smith and Martha Moutray, who offered unflagging professional support and encouraging words throughout the gestation of this work.

In the end, nothing of this book could have been conceived, executed, or completed without the support, understanding, and tolerance of my family, Marilyn, Kyle, and most recently, Cameron. "Love moves worlds" (Shakespeare).

I

The First Modern Planet

Since man, fragment of the universe, is governed by the same laws that preside over the heavens, it is by no means absurd to search there above for the themes of our lives, for those frigid sympathies that participate in our achievements as well as our blunderings.
—M. Yourcenar, *Mémoires d'Hadrien,* 1958

ENCOUNTER WITH A NEW UNIVERSE

*N*ot long ago, while browsing a secondhand bookstore, I came across a volume entitled *The New Atlas of the Universe,* written in 1988 by the well-known popularizer of astronomy, Patrick Moore. The title of this handsome work, I admit, took me aback. Could it be true that the entire cosmos had really been probed, explored, mapped—and updated? But the book turned out to be far less than this, and therefore, in many ways, far more interesting. It was, in fact, an atlas of the solar system (a somewhat provincial version of "the universe"), consisting mainly of detailed images and maps of the planets and their moons, along with respective lists of surface features recently identified by various spacecraft.

This might sound rather humdrum. Yet another view of Jupiter's giant red spot? One more close-up of Saturn's auroral rings? Mars, as we know it so well, still a rusty, windswept, and boulder-strewn surface? Such was the visual chorus I expected to find, a coda of images tantamount to photographic clichés. But I was in for a number of striking surprises. Leafing through the pages of this book, I found myself entering a "universe" I had no idea existed.

As a geologist, I had been generally aware of the visual riches culled from the two Voyager space probes launched by the United States in the late 1970s. These robot eyes sent out to wander among the worlds and satellites of Jupiter, Saturn, Uranus, and Neptune had reportedly brought back tales and wonders of these geographical *Novae Orbi* and had begun a new era of "contact." This I had known; but here was the overwhelming evidence, of which I had been ignorant. Here were images that revealed worlds of unaccountable feature and action. Here were the violently eruptive sulfur volcanoes on Io, spewing gases and ions far into space. Here were the eerie, spidery linea of Europa, stretching for hundreds of miles just beneath a glazed skin of frozen methane. Here, too, were the gigantic, broken ice cliffs on Miranda, rising to heights that dwarfed even the Himalayas; and the multiform terrain of Triton, whose patchwork landscapes seemed grafted onto each other without reason or order, as if by collision. Here were images, in short, that offered a particular inebriation.

To geologists, the Earth is huge and visually infinite; this is a conceptual necessity. With its innumerable subfields and levels of scale, the geological Terra is a universe all its own. The famous portrait of our planet as a single ball, swirling with cloud, taken by Apollo astronauts on their way to the Moon, is in no way a geological view. It is too distant, too complete, too unified—indeed, too much like the Moon itself. It is therefore something else: an aesthetic vision that has left the gravity of science behind.

Habituated by my own geological training and knowledge, I was never quite prepared upon opening this *New . . . Universe* to encounter the faces of so many worlds, dangling in the black of space, their features available to the eye of instant interpretation. Within this book, each planet and moon had its accompanying map, composed of a computer-generated image that flattened its subject out on a single rectangular strip—the so-called Mercator projection. This, too, seemed interesting: a technique literally four hundred years old invented at the height of the early colonial era, the Age of Exploration, now being employed to make visible the most advanced geographies in a new age of discovery. Indeed, what might Mercator have thought were it suggested to him that his scheme would one day be used to plot landscapes so far from terrestrial in aspect as to reflect back, in their magnificent alienness, the very idea of an old and exhausted Earth?

Other surprises were yet to come. Staring at these maps, pondering the contour of their features, I became aware that their alien qualities had been reduced in a particular way. They had been made, in a sense, part of Earth itself. The possession came in a simple form: names. Shifting focus to these, I found each world coming into view in an entirely different way. If it is true (as Goethe once said) that every act of naming is a birth, a mapping of culture and history onto the world of things, then here indeed was a revolution in the skies. Here was an end to the Old World of the heavens, titled after the legends of Greece and Rome and the heroes of Occidental astronomy.

On Mercury, for example, none of the craters bore names having anything to do with classical myth or famous scientists, nor were they written in Latin. On the contrary, these conventions had been entirely overthrown for the sake of a different canon altogether: the names of authors, painters, musicians, sculptors, and poets, not only from Western nations, but also from China, Japan, Islam, Africa, and India. Instead of Cassini, Kepler, or Laplace, one found Saikaku, Keats, and Li Ch'ing Chao. In the place of Olympus mons or Valles marineris there were Suisei planitia and Pourquoi-pas rupes. Something, it seemed—both of the "two cultures" conflict between the sciences and humanities and of Western self-infatuation—had been overcome on the lifeless and broken plains of this heated world. Venus—made visible by both the U.S. Pioneer and the former-USSR Venera 15 craft in the early 1980s—seemed to address other inequities. Earth's "sister planet" had become the abode of female heroines and deities, again from a wide variety of cultures: Devana, Sacajawea, Lakshmi, Freyja, Ishtar, and Niobe. The moons of Jupiter presented a diversity even far more remarkable: Io as the multicultural planet of fire, Sun, and thunder gods (male and female); Callisto, the world of Norse mythology; Ganymede, populated by deities of ancient Babylonia and Egypt; Europa, wanderland where Celtic and Greek characters meet. Circling Saturn were a host of literary sources: Mimas, whose surface bore the names of people and places in Malory's *Le Morte d'Arthur;* Enceladus, scene of the *Arabian Nights* (the Richard Burton translation); Iapetus, where the *Chanson de Roland* is now endlessly sung; Tethys and Dione, of the *Odyssey* and *Aeneid*. All Uranus is a Shakespearean stage, its satellites—Miranda, Ariel, Umbriel, Oberon, Puck, and Titania—dancing forth from *A Midsummer Night's Dream*.

A new universe, therefore, has truly opened up. A new postcolonial era in geography, projected into the skies, has begun. Within the domain of Western science, the heavens (in their nearest form) have at last become a true collection of voices from many lands, an ecumenical polyphony. They no longer belong exclusively to the Greco-Latin West.

A THOUSAND QUESTIONS

Names are an expressive finale to exploration. Attempts to penetrate the distant and the alien evoke the power to define "place" and lay claims upon it. But ideas of what constitutes a geographical feature or an entire landscape are tightly bound with mental images taken and augmented from the past. At a fundamental level, our "new universe" may not be so new after all, for it depends on a series of ancient concepts, of which "planet" is the most obvious. What precincts of the imagination helped fill this idea of "planet" with the possibility of landscape? Don't the raging volcanism on Io, the frozen seas of Neptune, or the poisonous rain and air on Venus recall ancient premonitions about the dangers and spectacles to be encountered beyond the reaches of the known world? For that matter, don't the endless presuppositions about alien forms of life on Mars, Venus, or wherever—both in scientific literature and in science fiction—resemble the accounts of fabulous creatures, monsters, or displaced souls so long attributed to distant terrestrial domains?

Any of these questions might be the topic for a rich historical inquiry into the contours of the Western imagination. In asking where the new worlds of today place us, at what type of epistemological and cultural center, we are also asking where we have been situated in the past. Names thus emerge as historical expressions of a long gestational process. Names have arrived very late in the earthly story of the planets. Moreover, as the discussion above suggests, they are infinitely richer and more interesting as the bearers of history than as mere nomenclatural flags or tombstones. Indeed, what emerges when one examines the new planetary naming schemes more closely is that the seeming diversity reduces down to a few canonizing styles that have selectively interpreted the "high culture" of other peoples according to tenets integral to the Western outlook.

The "new universe," in fact, turns out to be something quite old, even hallowed. Cast over a vast array of spectacular forms and events, over planetary landscapes never before seen, is a sensibility—nay, a way of seeing— hundreds, even thousands, of years old. Names are being selected for the indefinite future by a series of nomenclatural committees chosen by the International Astronomical Union (IAU), an official body mainly composed of members from Europe and North America, with fewer representatives from India, China, and other non-Western nations.[1] To think that this composition is the ultimate determining factor in nomenclatural decisions, however, would be short sighted and wrong. Rather, certain traditions that were already well established within modern astronomy in the seventeenth century are being carried forward, traditions created and concretized in the first true mapping enterprise aimed at another world—the Moon.

"Any map," writes R. A. Skelton, "is a precipitation of the spirit and practice of its time." This, however, "should not blind us to the essential continuity of the historical process by which [maps] have evolved and are still evolving. Even modern maps yield more readily to analysis and classification if they are considered the end products of a continuous evolutionary process, reaching back into the Middle Ages."[2] Until quite recently, maps were much larger documents than they are conceived of today. They were places where a profound merger took place among art, science, mathematics, politics, and religion. During the early colonial era, from the sixteenth to the eighteenth centuries, maps were nothing less than the grand documents of Western civilization. Into them were poured nearly every major domain of learning from geography to classical literature. And into these domains of scholarship were mixed the contents of private ambition, the religious strife of the Counter Reformation, colonial sensibilities about the New World, and the rise and fall of nation states. All of this, in one form or another, found its way onto the earliest maps of the lunar surface.

EARTHLY POSSESSIONS

If there are ironies to be found in the new naming schemes for the solar system, the most striking is surely this: that the apparent variety among all these schemes is actually less than that for the Moon, the first "other world." The Moon, it appears, was far more vulnerable in the seventeenth century to becoming an open document, stained with the historical realities of the age, than the planets collectively seem today. As the most dominant yet changeable element of the night sky, the Moon has set the pattern for planetary imaginings since the most ancient times. Certainly it has drawn to itself a greater range of imaginative energy than any other body, with the possible exception of the Sun. In a sense, all of the solar system shines with the conceptual and nomenclatural brilliance of the Moon. Any and all interest in the images and names we have given the planetary bodies must begin here, where the patterns of projecting the Earth were first tried, tested, and confirmed.

The names of the planets themselves were inherited by Europeans via Roman culture. The tradition of naming worlds after Latin versions of the Greek pantheon was maintained into the twentieth century with the discoveries of Uranus (1781, by Sir William Herschel), Neptune (1846, by Johann Galle), and Pluto (1930, by Clyde Tombaugh). Yet these conventions had little impact on the titles proposed for the lunar surface when it was first mapped between 1630 and 1650. These, instead, emerged out of a number of other traditions associated with the idea of the Moon as a mythological, allegorical, artistic, scientific, and even political entity. That the Moon should come to be a domain over which the ghosts of famous scientists, philosophers, translators, poets, and saints would one day wander and debate the primacy of ancients vs. moderns—a dead world planted with agricultural references, astrological clichés, and allusions to mountain systems of the Earth—was an outcome that could not have been easily predicted at any stage but that bears a burden of history perhaps surprising in depth.

One thing is certain: It cannot have been a coincidence that the lunar surface was first named during the same era when "new worlds" were still being discovered on the Earth. Indeed, the mapping of the Moon, which began even before Galileo turned his telescope heavenward in 1610, sought to expand something specific on the Earth, something of urban, seventeenth-century Europe, upward into the heavens. No less than the New World of

North and South America, the Moon proved to be contested territory: the naming systems applied to it by various astronomers in the early part of the Scientific Revolution, like those of colonial explorers to foreign domains, reflect different acts of attempted possession, hopes, and ambitions that melded institutional and personal canonizations.

In its geographic nomenclature, the Moon remains a document of a specific and critical moment in the evolution of modern Europe, and it is this moment, now hardened against time, that we encounter whenever we see a map of the lunar surface, whenever we look skyward at its forms with names in mind, or whenever we gaze at the "new universe" of planetary geographies that now surround us with Earthly projections. These names bring to the eye of the present examples of how the objects of scientific study are stamped deeply with cultural meanings and finally how such meanings, once frozen, become intrinsic to science thereafter.

THE LUNAR IMAGINATION

This is a book about the Moon and its place in the Western imagination. It is a book about art, science, literature, and to a certain degree, philosophy, inasmuch as all of these fields have had an abiding interest in the Moon and its nature. I make no pretense to being comprehensive in any of these areas but instead have sought to draw from each of them the evidence of how lunar reality has been conceived over the past several millennia, beginning with early Greek poetry and ending with the mapping of the lunar surface in the seventeenth century. At a broad level, the chapters that follow are concerned with imagery—the representations of lunar reality in written and pictorial form from the time of the earliest Greek poets down to the Scientific Revolution. On a more detailed level, however, these pages seek to reveal how conceptual traditions applied over time to a single object can both transform and stabilize that object in an ever changing dance of association that eventually is inducted into the realm of science.

My coverage of ideas and images regarding the lunar orb is not meant to be, and is far from being, inclusive in a cultural sense. I have dealt only slightly or not at all with the Moon as it has appeared in folklore, music, oral tradition, or religious ceremony. Instead, my intent has been to locate and trace through time the portion of lunar imagery that found its apotheosis in the concepts and naming schemes of the seventeenth century, particularly those that were placed upon the Moon's surface and that worked their influence upon subsequent notions of planetary reality.

In general terms, this book can be divided into three sections. Chapters 2–5 deal with images and concepts of the Moon as framed in antiquity and how these were later transferred to Europe in the Middle Ages, where they achieved textual, scientific, and artistic expression. Chapters 6–9 bring these trends into the Renaissance and the early years of the Scientific Revolution and discuss lunar imagery as it appears in art, literature, and the earliest observations by scientific authors. Finally, in chapters 10–14, the saga of how the Moon came to be mapped and named during the Age of Exploration, thus providing a model for the rest of the solar system and our modern concept of planetary reality, is treated in detail. This bare-bones outline is not meant to dissuade the reader from the hope for more wide-ranging contents. These, I believe, certainly exist in the many discussions of individual writers, texts, paintings, diagrams, and other ingredients in the grand tapestry of lunar associations. Indeed, if there is a single conceptual thread that weaves through the entire fabric of this multiform textile, it would be the notion that Copernicus, in some sense, was wrong after all: the Earth has never ceased being the center of the universe.

2

ℋow the ℳoon ℬegan

IDEAS OF LUNAR REALITY IN ANTIQUITY

The Moon is a phenomenon of the day and night sky that has been closely, lovingly, and fearfully observed by human beings of all cultures for thousands upon thousands of years. Long before Galileo and the first heavenward transport of the human eye by telescope, the lunar disk had gained a central role in the astronomy and astrology of peoples across the globe, not the least of which were those of ancient Greece and Rome. For the Greeks and Romans, who inherited large portions of their astronomy from the Babylonians and Egyptians, the Moon was the site of many earthly projections. Not merely a timekeeper with its phases or a mythological nexus, it was understood and perceived to be a source of realities such as weather, the harvest, illness, a broken psyche, fertility, and femininity. More than a resting place for the deposits of men, the lunar orb, as cultural object, had long been a site for moral, spiritual, philosophical, and literary presuppositions before Greek philosophers adopted it as a subject.

In addition to its size, changing aspect, and brightness, the Moon is unique in being the only celestial body with surface features visible to the unaided eye. In drier climates, such as those that have governed the Mediterranean region during the past several thousand years, these features present

*Figure 2.1. Babylonian clay tablet recovered at Seleucid Uruk and dated before 800 B.C.,
showing the seven stars of the Pleiades, the image of a bull (constellation Taurus), and the
Moon with incised figures. Reprinted by permission, Tafel VAT 7851, Staatliche Museen zu
Berlin Preussischer Kulturbesitz, Vorderasiatisches Museum.*

themselves all the more starkly and magnificently.[1] It is impossible that the
Greeks in particular did not perceive the lunar maria on a regular basis.
Indeed, there is much evidence in the textual remains we possess that they
did and that they applied to these "spots" a range of competing interpretations. Visual representations exist as well. Images of the lunar crescent appear
in various cosmological and astrological works of Egyptian and Babylonian
origin, for example, and much later in several illustrated versions of the *Iliad*
and *Aeneid.* It seems likely that these represent only a small remainder of a
much larger corpus.

In a few cases, we find glimpses of a truly complex style of visual imagination. The clay tablet of figure 2.1, for instance, is from the Babylonian site
at Seleucid Uruk, dated before 800 B.C., and shows the Moon as a circular
disk containing a bearded male figure who holds a club in one hand and an
animal by the tail in the other. The figure is thought to be Marduk, creator of
heaven, Earth, and the human race.[2] To the left are the seven stars of the
constellation Pleiades depicted in pointed fashion; to the right, we see part of
a bull, representing Taurus. The written text above these images refers to an
astrological prediction of a lunar eclipse in the month associated with the

zodiacal sign of Taurus (late April–May). What interests us here, however, is the lunar image. Is this an attempt to interpret the visible features of the lunar disk in terms of some divine story, perhaps a precursor to Herakles' defeat of the Nemean lion? No definite answer can be given, yet the suggestion lingers. Obviously, the Moon was a subject of detailed portrayal very early on. Even if no naturalistic images of the lunar face were produced before the threshold of the European Renaissance, the tradition of depicting a figure or face in the Moon's orb recurs, like a template, in both word and paint throughout the whole of the classical and medieval periods. This seems significant, if not exactly surprising: the discovery of humanlike forms embedded in the lunar surface begs that a connection be made to equally venerable assumptions regarding the inhabited nature of this surface, and from there, to the notion of life on other worlds.

VERSIONS OF THE MOON IN CLASSICAL ANTIQUITY: THE BEGINNINGS

Earliest Images

The earliest images of the Moon in antiquity are known to us only through secondary sources, all of which were written centuries after the nature philosophers and poets they claim to cite. With the exception of Hesiod perhaps, the actual words of these first thinkers are forever in doubt or lost.[3] If, however, these intermediary works are at all accurate, every one of the ancient poets and pre-Socratic philosophers made important statements about the nature and substance of the lunar orb, and these statements form our necessary starting point.[4]

The oldest Greek images of the Moon are cosmogonic ones, meaning they deal with various aspects of cosmic creation resulting from a diverse set of marriages and unions between personified deities. Hesiod (eighth century B.C.) in his *Homeric Hymns,* for example, speaks of "bright Selene, daughter of the lord Pallas, Megamedes' son" and portrays her as a goddess who reveals her "immortal head, [from which] a radiance . . . embraces earth." She bathes "her lovely body in the waters of Ocean, [dons] her far-gleaming raiment, and yoke[s] her strong-necked shining team." Above all, she is "a sure token and a sign to mortal men."[5] Similar images can be found in the works of Homer and other early poets, such as Musaeus, to whom is attributed the claim "I am the offspring of Selene of the beautiful hair, she who, seized with terrible shuddering, brought forth the Nemean lion."[6] The lunar body sometimes played an important role in myths related both to brightness or prophecy and to the darker side of existence. As deity, Selene (from *selas,* meaning "light") was invoked not only as the sister of Helios (the Sun) as a patron of the hunt and the virgins of epic tales, but also as a spiritual sister to Hecate, "daughter of darkness," and in league with the Medusa.[7]

Another side of lunar imagery viewed the Moon as an actual place, mythic in origin. It was associated with Elysium, the "islands of the blest," where, according to Homer and Hesiod, the gods sent favored heroes "far to the west," beyond the Pillars of Herakles (Gibraltar), that they might not suffer the indignity of death. Indeed, Elysium was used somewhat generally as a site of transmigration. According to one Homeric hymn, "The substance of the soul is left upon the Moon and retains certain vestiges and dreams of life."[8] Doubtless the most striking geographic images to be found at this early stage in lunar history appear in fragments from the ancient poet Orpheus, who stands at the border between oral and written culture. One such fragment, included among what are known as the "rhapsodies" *(rhapsodiai),* meaning "sacred parts,"[9] deals with the creation of the cosmos. Chronos, father of all, shapes an egg out of the ether, and from this egg springs Phanes, "firstborn" among gods. Phanes is himself built of many parts: golden wings, four eyes, the heads of different animals, the voices of a bull and a lion, male and female genitalia, and many names. He is the very icon of natural fecundity and power, an image of excess and superabundance. Overflowing with the forces of earthly creation, he becomes the builder of other realms, one in particular: "And he devised another world, immense, which the immortals call Selene, and the inhabitants of Earth Mene (both words mean Moon), a world which *has many mountains, many cities, many mansions*" [italics added].[10] Thus, at the very earliest point, we find the lunar surface cast as another Earth, an image that our own era has finally confirmed through actual visitation: the presence of "mountains," and by implication, plains, valleys, and other terrestrial-like features.

In the beginning, therefore, the Moon was many things: a personified deity, a source of inspiration or evil magic, an inhabited geography populated by unknown races or by purified or purgatorial spirits. As such, the lunar orb could be both a "lower" and a "higher" Earth, and this is precisely the polar position it retained for centuries thereafter.

The Pre-Socratics: Selected Imagery and Influences

It is common to look upon the pre-Socratic philosophers of the sixth and fifth centuries B.C. as the true originators of philosophy and science in the West. Yet the writings associated with this time period show that "no line is yet drawn between philosophy, theology, cosmogony and cosmology, astronomy, mathematics, biology and natural science in general."[11] Put differently, the seeds of more rational considerations of nature, especially the heavens, are clearly in evidence alongside (and sometimes within) the older, more poetic traditions.

The first pre-Socratic thinkers about whom we have any reported information are three sixth-century authors of Miletus, then a flourishing Ionian city in Asia Minor. It was Aristotle who later grouped these three into his so-called Milesian school, consisting of Thales, Anaximander, and Anaximenes. Thales, the earliest, was considered one of the "Seven Sages" of ancient Greece and is said to be the first to maintain that the Moon shines by reflected light from the Sun. This was reported by Aëtius in his work, *Doxagraphi graeci* (ca. third century B.C.), who repeated what others before him had written. Aëtius also stated, "he [Thales] says that eclipses of the Sun occur when the Moon passes directly in front of it; explaining that the Moon is *of an earthly nature,* even though it gives the appearance of a disc" [italics added].[12]

Thales, a Phoenician by birth said to have absorbed the astronomy of the Egyptians and the Babylonians, is given the first claim based on observational evidence that the Moon is, in fact, like another Earth. Such an image stands in contrast to that offered by Anaximander, Thales' pupil, who believed not only that the Moon shines by its own light but that "each of the heavenly bodies is a wheel of fire, surrounded by air. . . . The air has little breathing holes somewhat like the holes in a flute, and through them the orbs are seen. When the hole of the [solar or lunar] orb gets clogged, an eclipse occurs. The moon goes through its phases as its breathing hole gets successively opened and stopped up."[13]

Anaximenes, the third member of the Milesian school, proposed yet another version of lunar substance, derived from his belief in air as the primal cosmic material. The Sun, stars, planets, and Moon all have a similar origin: mist rising from the Earth condensed into fire, which in turn rarefied into the heavenly bodies, retaining its fiery substance by virtue of rapid movement. The Moon moves more slowly than the Sun and is therefore cooler, a flat disk floating in space.[14]

These three accounts of the lunar orb show that the Milesian thinkers were moving beyond the purely mythological-religious cosmologies of earlier centuries, while at the same time retaining something of past beliefs. The Moon is no longer personified as a goddess, nor is it a way station for spirits or daimons. Yet it continues to emerge from a primal fecundity that bears direct relation, and probable inheritance, from older Egyptian, Hebrew, and Babylonian creation stories. Although the hands of the gods may no longer be visible, their fingerprints remain. The Moon reflects back the face of a society searching for new narratives appropriate to a culture of slowly expanding literacy yet steeped in poetic storytelling. Physicalist explanations are those of prose more than of poetry. The Milesians appear to have inhabited a domain between these two end terrains.

Poetic Traditions

If Thales and his pupils represent the ascent of prose to the heavens, the poetic realm was no less well endowed with authors: for example, Heraclitus, Parmenides, and Empedocles. Heraclitus, known for his Zen-like parables and his concept of the universe as an infinite expression of change, is reported to have described the stars and planets as "bowls with their hollow side turned toward us . . . [such that] bright exhalations collect in these concavities, where they are vaporized into flame." Moreover, "The Sun's flame is the brightest and hottest of these; the other stars are farther away from the Earth, which is why we receive less light and heat from them. The Moon is nearer to the Earth, but it has to travel in a region that is impure. The Sun, on the other hand, moves in a region that is transparent and unmixed, which is why it gives us more heat and light."[15]

Such rendering of the Moon into a surface touched by notions of purity and impurity was shared by Parmenides, whose epic work *On Nature* (mid-fifth century) proposed a cosmic scheme in which the Sun and Moon were both formed of fiery matter taken from the Milky Way, with the Sun evolving into hot and subtle substance and the Moon into matter more primitive, dark, and cold.[16] The Moon reveals its mixture of dark matter and fiery substance in the black spots that cover its face, even as it "wanders around the Earth, shines at night with a light that is not her own . . . always gazing toward the rays of the Sun."[17]

Empedocles (early fifth century B.C.), one of the most often cited poets of the pre-Socratic period, offered a similar image. The Moon, said this poet, "was composed . . . out of the air cut off by the fire. For this air froze just like hail." The lunar body is a "broad disc" and produces eclipses by being a solid body.[18] Empedocles further spoke of the lunar body "turn[ing] like the nave of a chariot wheel round the goal at the extremity . . . gaz[ing] on the holy circle opposite of its master."[19] He seems—although again, as with Parmenides, the language is anything but clear and definitive—to have posed the Moon as orbiting the Earth and perhaps shining by light from the Sun.

Despite the rise of prose narratives about the heavens as well as the advent of mathematical astronomy, the poetic lunar tradition did not decline and disappear in the following centuries. Instead it underwent a significant change, perhaps in the fourth or third century, when literacy became much more widespread, the technology of making books more sophisticated, and the desire to own them more prevalent. For this new, popular audience, poetry on the heavens entered the realm of what we today would call "science popularization." Such a designation, of course, is inappropriate on any but the most general level; however, it gives some flavor as to the type of writing now practiced in this tradition.

By far the most famous and influential author in this genre was Aratus of Soli (early third century B.C.), who wrote in hexameter rhyme a descriptive epic of the constellations, their basic movements, and their astrological significance, all based on a much more difficult, technical work by the mathematical astronomer Eudoxus. *Phænomena* (a title Aratus adopted from Eudoxus's original) was not only the most popular astronomical work of its day, but also served as one of the primary sources of Roman astronomy, remaining widely read, studied, and relied upon as a basic reference until the European Renaissance. How did Aratus speak of the lunar orb? Here is an example:

Scan first the horns on either side the Moon. For with varying hue from time to time the evening paints her and of different shape are her horns at different times. . . . From them thou canst learn touching the month that is begun. If she is slender and clear about the third day, she heralds calm; if slender and very ruddy, wind; but if thick and with blunted horns she show but a feeble light on the third and fourth night, her beams are blunted by the south wind or imminent rain.[20]

It has been said that Aratus's fame stemmed from his ability to exploit a taste for nostalgia regarding the heavens, here evident in the image of the Moon (once again) as a personified sign of terrestrial events. Aratus does not seem to have been interested in providing a poetic version of Eudoxus's complex mathematical astronomy. *Phænomena* instead used scientific observations and a somewhat elevated poetic imagery to strengthen an already ancient heritage of semiastrological beliefs and expressions, current among the populace at large. Aratus states at the beginning of his work that his main task is "to tell the stars"—not, we might note, to explain or portray them. The enormous success of this work is partly allied with its playful and ornate aestheticism and partly with its power to legitimize this very view of the heavens. In Aratus's hands, the lunar body is a nexus for a crucial mixing of literary, mystical, and scientific images that reveal the varied role the stars and planets continued to play in Greek society.

Plato and the Moon: A Conservative Theology

In *Apologia,* Plato has Socrates call the idea of an Earth-like Moon "absurd." But elsewhere in Plato's writings, such a notion is deemed much worse: an irreverent or even "dangerous" belief. In large part, this is because of Plato's concept that the planets were created as "living creatures" and remained the material expression of the one "world-soul," an idea common among the Pythagoreans. Plato expresses it in several works in several ways. In *Timaeus,* for example—his work that had the most influence on astronomy through the Middle Ages—the origin of the solar system takes place this way:

The Sun and Moon and five other stars, which are called the planets, were created by Him in order to distinguish and preserve the numbers of time, and when He had made their several bodies, He placed them in [their] orbits. . . . Now when each of the stars which were necessary to the creation of time had come to its proper orbit, and they had become living creatures having bodies fastened by vital chains, and learned their appointed task . . . they revolved, some in a larger and some in a lesser orbit.[21]

Elsewhere, Plato mentions "the deity or divinity of Sun and Moon . . . those souls good with perfect goodness [that] have proved to be the causes of all,"[22] and he condemns outright as "mischievous" and "dreadful" "the theories of our modern men of enlightenment" who preach that the planets, including the Moon, "are but earth and stones."[23] These men, says Plato, "are made infidels by their astronomy," for, by seeing the heavens as merely a collection of "soulless bodies," they ignore the primal intelligence and divine order in all creation.[24] Yet there are still hints that the planets are habitable bodies: "Having given all these laws to His creatures . . . the creator sowed some of them in the Earth, and some in the Moon, and some in the other instruments of time [the planets]."[25] Plato clearly carries forward the spirit-populated idea of the Moon that had existed for centuries, since the Orphic period. If we cannot rule out Plato as a potential supporter of extraterrestrial (lunar) life, this is because of his own brand of poetic conventionalism.

Although *Timaeus* is infused with "soul," it also deals with the geometry and mathematics of the solar system, albeit in rudimentary fashion. Plato's fundamental view of the universe returns to a formal perfection:

And [the Deity] gave the universe the figure which is proper and natural. For the living thing which should contain within itself all living things, [only] that figure would be proper which contains all other figures. Thus He made the world in the form of a sphere . . . having its extremes in all directions equidistant from the center, the most perfect and the most like itself of all figures. . . . This He finished off with a surface perfectly finished and smooth.[26]

A spherical universe containing spherical stars and planets had been proposed before, most notably by Parmenides and Pythagoras. But Plato gives the image an added quality by placing it at the center of his own conservative moral theology of the cosmos.

From this point on, the Moon and all the other heavenly orbs gain a new image—that of a smooth and polished globe, as perfect and complete as the Creator himself. Although at least one earlier author (Ion of Chios, late fifth century B.C.) had proposed a "transparent, glass-like material"[27] for the lunar substance, Plato was the one to attach this notion to that of the sphere and the sphere itself to the image of the pure, eternal, and divine order. Plato thus stands somewhere between the older, oral-poetic tradition steeped in mythic images of purity and a newer, rising realm of interpretation employing the power of written prose and mathematics. In the realm of celestial imagery, Plato stands as a bridge between cosmogony and astronomy.

The Moon as Another Earth: A Tradition Reestablished

The idea first attributed to Orpheus and Thales, that the Moon is of an "earthly nature," gained a solid and more literal footing among two different groups of later pre-Socratic thinkers. One group comprised the followers of Anaxagoras (mid-fifth century B.C.), known as the first great "teacher" of Athens, and his disciple, Democritus (late fifth century), leading proponent of atomism. For Anaxagoras, "the Moon, like the Sun and stars, is an 'incandescent stone'. . . [and] its 'turnings' have the same cause as the Sun's, but are more frequent 'because it cannot master the cold.' It is of earthly substance, though with some fire in it, and contains plains, mountains, and valleys."[28] The origin of the Moon is the Earth itself. Anaxagoras had known of the fall of a sizable meteor in 467 B.C. at Aegospotami. From this event he seems to have derived the idea that stonelike masses were originally solidified out of the Earth and hurled into the ether, there to become the rotating planets, stars, and other bodies circulating around the Earth.[29]

There has been much debate on whether Anaxagoras considered the Moon to be inhabited. He appears to have hinted that life did exist there; following the poet Orpheus, he says the Nemean lion fell from it. But his thinking went further than this, posing in enigmatic terms one of the first formulations of the plurality of worlds: "Men were formed [on other planets?], and the other creatures which have life; the men too have inhabited cities and cultivated fields as with us; they have also a Sun and a Moon and the rest, as with us, and their earth produces for them many things of various kinds, the best of which they gather together into their dwellings and live upon."[30] For such ideas, Anaxagoras was banished from Athens, being possibly saved from execution by his friend, Pericles. Yet this did not prevent Democritus from espousing similar notions only a few decades later, and going still further by openly maintaining both the existence of a plurality of worlds and an interpretation of the dark areas on the lunar surface as shadows cast by hills.

Nor did the treatment accorded Anaxagoras keep such images from becoming widespread and well rooted, even in Athens, during the following century. They even appear in Plato's *Apologia*, where Socrates seeks to defend himself against accusations of blasphemy by refuting the charge that he believes "the Sun is a stone and the Moon a mass of earth." To this allegation, leveled by Meletus, Socrates responds: "Do you imagine that you are prosecuting Anaxagoras, my dear Meletus? Have you so poor an opinion of these gentlemen [the jury], and do you assume them to be so illiterate as not to know that the writings of Anaxagoras of Clazomenae are full of theories like these? And do you seriously suggest that it is from me that the young get these ideas, when they can buy them on occasion in the market place for a drachma at most?"[31] Anaxagoras thus commanded no small fame by this time (late fifth–early fourth century B.C.), and his ideas had probably been a source of controversy for decades. If what Socrates says here is accurate, we can assume that any religious objections to considering the Moon another Earth were by this time trivial or relatively powerless.

This also would have been true elsewhere in the Greek world, as indicated by Diogenes of Apollonia (later half of the fifth century B.C.), whose home lay in one of the Milesian colonies along the southern Black Sea coast. Diogenes had read Anaxagoras's works, being particularly impressed by the story of the fallen meteor at Aegospotami, and seems to have made a pilgrimage to the site, there noting a strong similarity between the pocked surface of the meteor and that of pumice. On the basis of this field observation, Diogenes thereafter considered the Moon, Sun, and stars to be made of the latter material, with fire from the outer ether passing through the pores of this celestial pumice and giving rise to the light of each body.[32] Diogenes is thus the first to propose a link between the Moon and volcanoes—a connection that would recur at various times down to the nineteenth century, by authors no less important than Kepler, Hevelius, Herschel, and Schröter. Had the Greek philosopher had access to a telescope, he would have found in the lunar craters and broad maria—indeed in the entire rugged and pock-marked appearance of the lunar surface—evidence in abundance for such a link. Diogenes offered one of the only images in antiquity that has been borne out by contemporary science.

The second group of Moon-as-Earth believers were the Pythagoreans. Here we return to the image of Elysium and the "isles of the blest," which Pythagoras associated with the Sun and Moon in his canon of maxims. So important was this image to the Pythagorean school that its later members came to consider Pythagoras himself as one the souls or daimones who lived upon the Moon in eternal perfection.

The notion of a more physically inhabited lunar surface, on the other hand, was expressed in the work of Philolaus of Croton (later fifth century B.C.), the first of the Pythagoreans to have produced written texts and there-fore known as a principal astronomical theorist for the school. Philolaus was the first thinker to propose that the Earth, Sun, Moon, and stars all orbit an invisible "central fire."[33] Unique in being neither geocentric nor heliocen-tric, this scheme nonetheless pales in its remarkable aspect when compared with Philolaus's view of the Moon. Simplicius reports this as follows: "He ex-plains the Earth-like appearance of the Moon by saying that it is inhabited like our own, with living creatures and plants that are bigger and more beau-tiful than ours. Indeed, the animals on it are fifteen times as powerful and do not excrete, and the day is correspondingly [i.e., fifteen Earth days] long."[34]

How far we seem here from the lunar fields of poetry in a Plato or an Empedocles! Yet not so far, perhaps. Philolaus enshrines the idea of the lunar surface as a setting for a *better* Earth. With time the Moon-as-another-Earth image has gained new dimensions, being first equipped with mountains and mansions (Orpheus) and now with superior plants and cleaner animals as well. Philolaus may not have been the only one to propose this idea. About the same time, a secondary poet, Herodorus of Heraclea, wrote a work on the myths surrounding Herakles in which he stated that "the women of the Moon are oviparous and those born there are fifteen times our size."[35] It may be that this author and Philolaus knew something of each other, or of a third, unknown author. Either way, the ancient association of the Moon and fecun-dity is here revitalized.

Later Greek Astronomy: The Mathematical Tradition

The tremendous expansion of Greek mathematical astronomy in the later fifth and fourth centuries B.C. was mostly predicated upon efforts to solve the problem of planetary motion. Accounting for the irregularity of this motion constituted one part of a twofold riddle for early explanations of the heavens. The other part concerned eclipses and phases of the Moon. As might be expected, this second part of the problem was essentially solved first by the conclusion that the Moon was a solid object and shone by reflected light. This was a substantial advance, and it had immediate implications for lunar imagery in general. Explaining the irregular paths of the planets as they wandered through the zodiac proved far more difficult to tackle and eventually gave rise to highly intricate, ingenious systems of mathematical description. What made these systems possible was one of the major intellectual movements of the time—the application of geometry to physics and astronomy.[36] This meant a profound change from speculative and poetic ideas to fairly rigorous systems focused almost exclusively upon the geometric forms of celestial motion, proportions of distance, and relative size.

The systems so conceived were, from the very beginning, highly complex and grew only more so with time. Their patrimony lay in several places: Parmenides and Plato, proponents of spherical conceptions; Philolaus and the Pythagoreans, who imbued the cosmos with number; and Babylonian and Egyptian mathematics, which had developed early forms of arithmetic analysis for planetary position. Plato's emphasis on the sphere as the crucial figure of "uniform and orderly" motion essentially dropped the heavens into the lap of the mathematicians. Certainly the spherical image captured the imagination of mathematically inclined thinkers, and they set about trying to find ways to make the universe conform to this image of perfect order. That these attempts grew ever more strained and manneristic in complexity need not concern us here, nor should the details of any of the relevant systems. What matters most with regard to the Moon and the other planets is how these new schemes supported some images of the lunar body over others. Two systems in particular are worthy of note.

The first is that of Eudoxus, a mathematician and teacher by trade, who had studied with Plato and who conceived a design for planetary motion that involved a series of concentric spheres. Eudoxus's model was an elaboration of several earlier geometric descriptions and was quite sophisticated in its details: the movement of any single planet was controlled by several spheres spinning in different directions, with a total of twenty-seven spheres for the entire solar system centered upon an immobile Earth.[37] This system had considerable influence, providing a standard on which all future advances and elaborations were dependent. Eudoxus's model was first expanded by Callippus in the late fourth century, and shortly thereafter, by Aristotle himself, who increased the total number of spheres to fifty-five. It was countered in the early third century B.C. by Aristarchus, a scholar at the Library of Alexandria, who claimed that the Earth rotated (an idea propounded by Heraclides a century earlier) and the Sun lay at the center of the cosmos. Aristarchus, however, found few followers. His system was largely forgotten under the crush of fame granted Hipparchus (mid–late second century B.C.).

Hipparchus put the Earth back at the center and greatly refined the idea of epicycles (small orbits centered on the circumferences of larger orbits) and eccentric orbits earlier proposed by Apollonious of Perga. The Hipparchian model was an extension of all that had gone before, being the first to employ in its analysis the techniques of trigonometry. It was a model that stood the test of time until receiving refinements under the embellished synthesis of Ptolemy (second century A.D.) set forth in *Syntaxis mathematica,* or, as it is better known through the Arabs, *Almagest.* Ptolemy's effort remains one of the grandest achievements of ancient astronomy, an unparalleled blend of mathematical and observational effort. Yet conceptually speaking, it never leaves the ground set down by Eudoxus. No less than his predecessors, Ptolemy sought to account for irregular motion in terms of uniform, circular movements. In a sense, *Almagest* represents centuries of elaboration upon this single theme, immobile at the center of Greek astronomy. If, as one scholar has written, *Almagest* "caused an almost total obliteration of the prehistory of the Ptolemaic astronomy,"[38] it did so by embodying this history, not by surpassing it.

In terms of imagery, the mathematical systems of Eudoxus, Hipparchus, and Ptolemy had the effect of draining the heavens of substance. The geometric narratives or "fictions" devised to explain planetary motion had no interest in the physical, material reality of the planets. The Moon especially was now often reduced to a point on an epicycle, a dot carried around the Earth.

Did Hellenistic astronomers use actual diagrams in their work? Almost certainly they did. The type of path described by Eudoxus as representing the

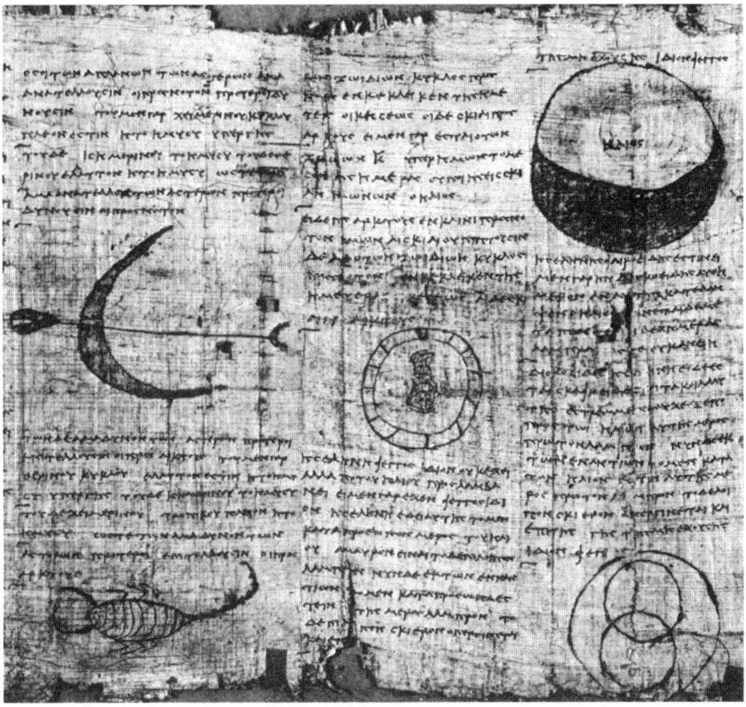

Figure 2.2. Portion of an introductory Hellenistic astronomical manuscript on papyrus, dated early second century B.C., *showing crude geometrical diagrams of celestial phenomena. Most of these diagrams bear little or no discernible relationship to the accompanying text. Reprinted by permission of the Louvre Museum, Letronne Papyrus, I.2.325. Photograph © Réunion des Musées Nationaux.*

actual motion of the planets, for example—a "hippopede," a kind of lightly deformed figure eight—surely required illustration of some sort. Euclid's *Elements of Geometry,* with its abundance of illustration, appeared at the end of the fourth century B.C., and a sizable portion of it (books V, VI, and XII) was apparently derived from Eudoxus.[39] Eudoxus is also well known as the first to have constructed a celestial globe, with the constellations pictured from the outside.[40] Aristotle, Hipparchus, Aristarchus, and of course, Ptolemy are all reported to have used globes, maps, and geometric diagrams.[41] Indeed, the oldest illustrated Greek papyrus in existence, written in the early second century B.C. and apparently intended as an introductory astronomical text, contains geometrical drawings depicting different aspects of the heavens (fig. 2.2).[42] These images are decidedly crude in execution and are difficult to decipher precisely because they bear only a partial relationship to the accompanying text.

A large circular drawing in the upper right of figure 2.2 is one such example. Labeled "Sun" *(helios)* in its lighter half, and "Moon" *(selene)* in its darker, crescent-shaped portion, it may be a literal restatement of the preceding lines of text, which state that the Moon has no light of its own but instead shines with the Sun's rays, because if the reverse were true, the far side of the lunar body would be luminous and the near side dark. The drawing may attempt to show that the lighted portion of the lunar surface "belongs" to the Sun. A portion of the text of this papyrus is an acrostic—a type of geometrical figure in linguistic form—spelling out *Eudoxus techne,* "the work of Eudoxus." This early scientific illustration is another example of a transitional creation: a new type of visual expression exists alongside, but is scarcely integrated with, much older traditions of purely linguistic descriptions.

Aristotle: A Moon of Enigmatic Substance

Next to Ptolemy, the most influential classical thinker with regard to the heavens was Aristotle. But where Ptolemy was concerned with the mathematics of celestial bodies, Aristotle's realm was the physical universe and its description in prose. Aristotle's *De caelo* (On the heavens) became a standard reference for more than a millennium and a half, gaining intellectual allegiance in Hellenistic, Arabic, and late medieval European culture down to the time of Galileo. What does this work say about the composition of the planets? We find this expressed in chapter VII of Book 2:

> The most logical and consistent hypothesis is to make each star [planet] consist of the body in which it moves, since we have maintained that there is a body whose nature it is to move in a circle. . . . The heat and light which [these bodies] emit are engendered as the air is chafed by their movement. It is the nature of movement to ignite even wood and stone and iron. . . . But the [planets] are carried each one in its sphere; hence they do not catch fire themselves. . . . Let this suffice for the point that the stars are neither made of fire nor move in fire.[43]

The logic here describes a circle of its own, perhaps reminding one of the mythical snake that swallowed its own tail. Aristotle was no astronomer, although he seems to have largely shared the astronomer's interest in motion and position as the central question of the heavens. Yet his planets are otherwise enigmatic things, and this is no less true for the Moon. In a work on animals, for example, he assumes that there may be unknown species that live upon the lunar surface, and he even speaks indirectly of the possibility of "men in the Moon."[44] This, of course, implies a solid substrate much like the Earth. But in another work, *Meteorologica,* one finds this: "We maintain that the celestial region as far down as the Moon is occupied by a body which is different from air and from fire . . . and is not uniform in quality, especially when it borders on the air and the terrestrial region. Now this primary substance and the bodies set in it as they move in a circle ignite and dissolve by their motion that part of the lower region which is closest to them and generate heat therein" (I, III).[45]

Here again, there is no specific statement of what makes up the planetary substance. Aristotle goes on to say that below the movement of the heavens, there is matter with qualities of hot, cold, wet, and dry: what is heavy and cold (earth and water) separates out and sinks to lower levels, whereas hot and dry matter (air and light) exists above.

Where does the Moon fit into this scheme? We are told only that the Sun's motion is rapid and near (because it produces both light and heat), whereas that of the Moon is slow and near (producing only light). We can conclude from this, on the basis of the passage cited above, that the lunar body is composed of a mixture of heavy/cold and hot/dry matter and that it is "impure" because it borders immediately on the terrestrial region. But more than this we cannot say. The Moon remains a borderland of unspecified reality.

The Aristotelian "Moon" meant different things in different contexts, and these various moons are expressive of the different versions that existed in Greek intellectual culture generally. When the topic was celestial motion, the Moon of the astronomers stood forth, spinning and massless. When the subject changed to the Earth and its weather, a lunar intermediary between celestial and terrestrial bodies was invoked. And when the concern shifted to the variety of animal life, to distant places beyond the "known world," another Moon emerged, a landing site for alternative existences. Aristotle's influence in these and other matters related to astronomy was profound and long lasting but at the same time contained. His work drew from, but left no marks on, the thought of mathematical astronomy. In truth, these two realms were kept apart for very specific intellectual reasons related to the classification of knowledge. This is shown by a passage attributed to the first century B.C. writer Geminus, whose work *The Elements of Astronomy* enjoyed no small success in drawing the limits of this particular discipline:

> It is the business of physics {*sic*} to consider the substance of the heavens and the stars, their force and quality, their coming into being, and their destruction. . . . Astronomy, on the other hand, does not attempt to speak of anything of this kind, but proves the arrangement of the heavenly bodies . . . and, further, it tells us of the shapes and sizes and distances of the earth, Sun, and moon, and of eclipses and conjunctions of the stars, as well as of the quality and extent of their movements. . . . The things, then, of which astronomy alone claims to give an account it is able to establish by means of arithmetic and geometry.[46]

Mathematics was thus seen as the intellectual fate of astronomy, its de jure discourse. What if the physicist and the astronomer converge upon the same phenomenon? "The physicist," says Geminus, "will prove each fact by considerations of essence . . . or of coming-into-being and change; the astronomer will prove them by the properties of figures or magnitudes, or by the amount of movement and the time that is appropriate to it."[47] There seems little room in this scheme for discussions of planetary substance.

THE EARLY IMAGERY OF ASTRONOMY

Prior to the time of Eudoxus, the language used to speak about the planets and the Moon was densely visual. Fiery stones, an inverted bowl, a glassy sphere, frozen hail, a chunk of pumice, another Earth ripe with mountains and cities: such were images of everyday experience, an empirical pantheon. The introduction of geometry produced a radical change. From the very beginning, it created a surplus of complex visualization: dozens of concentric spheres attached at the poles, spinning in separate directions; orbits encased and surrounded by other orbits, whirling at various speeds. From Eudoxus onward, words alone became insufficient to adequately describe and explain such conceptual schemes. The solar system required some form of pictorial rhetoric[48] to be made fully comprehensible. Mathematizing the heavens resulted in the first true examples of scientific illustration—the first consistent use of technical diagrams as a parallel discourse of clarification and summary. What type of discourse was this? As far as one can tell on the basis of meager evidence, it was largely or wholly Euclidean—a language of lines, circles, triangles, and rectangles; an extension of number itself, schematic and idealized, with no interest in representing actual material bodies.

It would be wrong, however, to say that the lunar surface became ever more absent as a subject for reflection. On the contrary, as we will see in the next section, the richness of the Moon as a possible geography continued to grow, becoming the topic of heated debate, discussion, even diatribe. For the most part, this took place not among astronomers, but within literature and philosophy. The lunar body remained a substantive reality outside the main trends of what today we would call "scientific thought." Science would one day catch up with and far surpass the material imagery of the literary imagination, but that day was still far off. Well into the Middle Ages, the poetic, novelistic, and philosophic traditions that had given birth to such imagery remained purveyors of the lunar surface.

3

Epic Journeys and Flights of Fancy

IMAGES OF THE MOON IN PLUTARCH AND LUCIAN

THE MOON AND THE ORIGINS
OF CLASSICAL GEOGRAPHY

Given the traditions of imagination that bound them, the lunar imagination and classical geography were fated to intersect. This they did, often and in several ways. Some of these ways have already been hinted at: the poet Orpheus spoke of lunar mountains and cities, and at least one branch of the Pythagoreans identified the lunar body as an "isle of the blest" to which souls and spirits migrated. Once it appeared, *geographia,* as the "writing of the Earth" *(geo, graph),* also depended upon literary and religious images, concerned as it was with the telling of tales of far-off lands.[1] Its version of the world was expressed in marvelous and frightening stories, stories of monstrous races and riches beyond measure, where the borders of the ordinary were surpassed or dissolved in spectacular fashion. Yet *geographia* did not cease with the further reaches of the terrestrial realm. There existed a continuum of fantasy between the distant Earth and the proximal Moon, a type of literary land bridge crossed whenever the margins of the present world (and credibility) were reached. Both realms were the subject of *historiae*—

flights into the remarkable, where the known often shook hands with its inverted somber or smirking self.

One of the most interesting and influential examples of such writing is the genre of the "miracle letter," a largely fictional form that claimed to reveal the marvels experienced by Alexander the Great during his campaign to "the eastern edges of the world." Such letters form a central part of *The Alexander Romance,* a loose novel probably written within a century of the young king's death. The longest of the letters in this work, and in many ways the most interesting, is *Alexander's Letter to Aristotle.* What makes this interesting is the representation of Alexander not merely as a conqueror but also as "a champion of Hellenic science" and "Aristotle's collecting arm."[2] What Alexander collects, however, are not specimens but fantastic adventures and sights, "observations" in short. The farther east he goes, the more monstrous and unfathomable these sights become, until, at the outermost limit of his travels, a group of sages brings him to a final oracle—two trees in the midst of a garden, one of which speaks with a male voice and is called "Sun," the other a female voice, "Moon."[3] At this moment, Alexander is given foreknowledge of his own death and that of his family. His eastward push is at an end; his implicit "quest for divinity" has stopped at the threshold of the lunar frontier.

The lore enveloping Alexander the Great was immense in later antiquity and constantly invoked the image of "other worlds." To the Romans, this image returned to the idea of empire and conquest. Regions such as Britain, Scandinavia, and Africa south of Egypt were all given the designation *alii orbi* ("other worlds") at one time or another. Continents far to the south, east, and north were called the domain of the antipodes or, more to the point, "counter-worlds," *antichthones.*[4] This was the very term that was sometimes used by the Pythagoreans to denote the Moon.[5] Such terms were not wholly reserved for writers of philosophy and fiction. In his work *Geographia,* Ptolemy adopted the name "Mountains of the Moon" to refer to the unreachable, snowcapped ranges that sourced the Nile.

Time and again, as ancient geographic writings approached the idea of the "ends of the Earth," the lunar body made its appearance. This is even better shown in the work *Wonders beyond Thule,* presumably by the second-century A.D. author Antonius Diogenes. In this story (actually a complex and tangled weave of stories within stories), spectacles and curiosities struggle to outdo one other as the setting gets farther from known regions. In Iberia, the protagonists encounter "a city where the people could see in the dark but were blind by day,"[6] a state later imposed on the heroes by evil magic, forcing them to lie covered in tombs during the day and emerge only in Moonlight.[7] One of the travelers, Astraois, is later granted great honor "because of the way his eyes grow larger and smaller indicating the waxing and waning of the Moon." The main heroine, Derkyllis, is captured at one point but finds solace in meeting again her brother, "who has wandered over many places and seen many marvelous things . . . about the Sun itself and the Moon and planets and islands."[8] Finally, "most incredible of all," the narrator "claims that in his voyage north he approached the Moon, as if towards some exceedingly pure version of Earth, and there saw the kinds of things you might expect from such a fictionalizer of outlandish fables."[9]

PLUTARCH'S *THE FACE IN THE MOON*: LITERARY SCIENCE AT ITS HEIGHT

Plutarch's dialogue on the lunar surface is one of the most famous and often cited works of classical astronomy. It is not, however, an astronomical book. It is instead a literary discussion, which embarks upon scientific, philosophical, and even mystical themes but does not always distinguish among them. This is clear even from the title, *On the Face Which Appears in the Orb of the Moon,* which adopts the oldest anthropomorphic image or metaphor associated with the lunar body. Plutarch has often been said to treat in this work all the major views of the Moon in classical antiquity. Indeed, the dialogue opens with one speaker requesting "those opinions concerning the face of the Moon which are current and on the lips of everyone."[10] Plutarch has thus been used as an occasional encyclopedia in the history of lunar studies.[11] The truth, however, is that the author has included only a select number of ideas with a particular purpose in mind. To begin any history of the Moon in Western thought with Plutarch would be an error, however efficient.

Narrative Structure and Basic Aspects

The dialogue takes place between eight speakers. There are two mathematicians, Menelaus and Apollonides, the latter a geometrician; a Stoic, Pharnaces; an Aristotelian, subtly named Aristotle; a proponent of the Academic position, Lucius; and a literary authority, Theon. In addition, Lamprias, the voice of critical reason sharpened by satirical wit, speaks for the author and has the major voice. Finally, there is also a traveler and teller of tales, Sulla, whose narration at the beginning and end encloses the dialogue. Nearly all branches of knowledge are therefore brought to bear upon the subject of the Moon. Plutarch is presenting himself as a virtuoso (which he certainly was). The Moon, he suggests, has attracted the imagination in every field of intellectual endeavor. But—and here is the deeper message—most of these efforts have failed to accurately comprehend this body and have made the attempt with vainglorious intentions. It is not, we learn, a mathematician or philosopher who possesses the lunar truth, but Lamprias, gentleman scholar of wide learning and literary leanings.

Sulla begins the dialogue near the finish of an apparently missing monologue. He has been relating a story he has heard, but before going further, he asks if it would not be better to discuss first the current views regarding the Moon's face because (it is implied) this is where the discussion will lead anyway. Lamprias responds that this is indeed necessary because it is always a good idea "to chant over ourselves the charms of the ancients and use every means to bring the truth to test" (p. 35). From here, Lamprias directs the talk to consider and then discount several theories about the nature of the lunar surface. These are, in order, (1) the idea that the "figure seen in the Moon" is an optical illusion; (2) the belief that the Moon is a mirror and displays a reflection of "the great ocean"; (3) the Aristotelian notion that the lunar body is an "ethereal and luminiferous star"; (4) the Stoic concept that the Moon is an impure mixture of air and "gentle fire" because it is without heat; and (5) the Academic position that the lunar body is glassy or icelike and that the light of heaven shines through it.

The discussion of these notions and disproof of them take up the majority of the book. Lamprias then suddenly recalls the original stimulus for the conversation: "[I]t is high time to summon Sulla or rather to demand his narrative. . . . So, if it is agreeable, let us stop our promenade and sit down upon the benches, that we may provide him with a settled audience" (p. 157). At this point, Theon interrupts to ask about the habitability of the Moon and the beings that are said to dwell there. The references that Theon mentions to press the topic are all literary or mythic: Aeschylus, Homer,

Hecate, and Athena. Lamprias responds in kind, maintaining that nothing has proven that the Moon is unpopulated by beings, animals, and plants. This possibility, moreover, is itself proven by the Earth, he says, which is a fount of life although it contains vast barren spaces. During his monologue, Lamprias seeks to outdo Theon in demonstrating, with all due verbosity, his own familiarity with classic literary sources, piling reference upon reference until, in apparent good-humored exasperation, Sulla cuts him off and finally launches his narrative with a quotation from the *Iliad:* "An isle, Ogygia, lies far out at sea" (p. 181).

The reader is thus transported to the ends of the Earth, to "a run of five days off from Britain," where there exists an island on which "Cronus is confined by Zeus" (p. 181). We have been prepared for this tumble into myth by Lamprias's own windy disquisition. His speech moved us backward in time, from Aristarchus, Plato, and Theophrastus to Alcman, Hesiod, and finally Homer. (Sulla's quotation is an obvious continuation of Lamprias's final two quotations from the *Iliad.*) Sulla's tale is secondhand; he is relating what some "stranger" told him, one who had returned from a "great mainland surrounded by ocean," where the natives worship the Moon. This stranger is obviously acquainted with ancient myth because he tells Sulla, "The substance of the soul is left upon the Moon and retains certain vestiges and dreams of life." (p. 215). Little of this purgatorial geography, however, can prepare us for what comes next:

> [J]ust as our earth contains gulfs that are deep and extensive, one here pouring in towards us through the Pillars of Herakles and outside the Caspian and the Red Sea with its gulfs, so those features are depths and hollows of the Moon. The largest of them is called "Hecate's Recess," where the souls suffer and exact penalties for whatever they have endured or committed after having already become spirits; and the two long ones are called "the Gates," for through them pass the souls now to the side of the Moon that faces heaven and now back to the side that faces Earth. The side of the Moon towards heaven is named "Elysian plain," the hither side, "House of counter-terrestrial Persephone." (pp. 209–211)

In these lines we see the very first evidence of an effort to *name* some of the visual features on the lunar surface. What is here called "Hecate's Recess" seems likely to be Oceanus Procellarum; the "Gates," on the other hand, may well correspond to the Mare Serenitatis, Mare Tranquillitatis, and Mare Fecunditatis.

This first naming scheme is closely tied to myth, yet at the same time, the "gulfs" on the Moon are given their direct, physical counterparts on Earth. Such analogies must have been noted previously and set down. Plutarch appears to be citing an earlier work that discussed these (and possibly other) names in more detail. What this work might have been we do not know. But Plutarch's own use of it had millennial consequences. Nearly fifteen hundred years later, the first European scientist to observe and draw the Moon by means of the telescope also noted a feature worthy of the title "Caspian."

In its physical aspects, the story told by Sulla therefore agrees with the major point of the dialogue. Plutarch, in the guise of Lamprias, stated this previously in no uncertain terms: "It is in fact not incredible or wonderful that the Moon, if she has nothing corrupted or slimy in her but garners pure light from heaven . . . has got open regions of marvelous beauty and mountains flaming bright and has zones of royal purpose with gold and silver not scattered in her depths but bursting forth in abundance on the plains or openly visible on the smooth heights" (p. 141). These "mountains flaming bright" were a legacy that Plutarch was to leave to the Middle Ages, which ignored it, to the Renaissance, which passed it by, and to the seventeenth century, which recovered it and gave it material truth in the form of drawn images and the first lunar maps.

Absences

Lamprias, with the occasional help of Lucius (who rejects the Aristotelian position), dispatches all of the mentioned theories about the Moon in fairly short order before pressing his own claim to the truth (from which we have just quoted). This claim sits upon the shoulders of poets: Homer and Pindar, Parmenides and Empedocles. The closest Plutarch comes to citing a scientific author is his reference to Anaxagoras and Plato.

This absence is matched by another. During the course of the dialogue, Lamprias reveals (as previously quoted) that the speakers have been walking in a garden and eventually sit down to pursue their discussion further. This is the only mention of scenery. Such was the centuries-old convention for this genre of writing. In this case, however, there is a difference. Instead of the typical philosophical or moral subject, a wholly visual, visible object stands at the center of the discussion. This object, moreover, is described throughout the work in terms of the features it presents to the observant eye. It would have been easy for the party to look up and observe the lunar face directly overhead—Aristotle would almost certainly have advised as much. The Moon itself is absent from the dialogue. For Plutarch, as for Lamprias, it is the subject of a literary science above all.

LUCIAN'S FABLES: THE TRUTH ABOUT HISTORY

A True Story

A century after Plutarch, the Moon and the distant journey were united once more, this time in a fantastic voyage with no pretensions to factuality or astronomical interest, literary or otherwise. Lucian, in fact, begins his fiction with an excess of qualifiers:

> Students, I think, after much reading of serious works . . . will find [this story] enticing not only for the novelty of its subject, for the humor of its plan, and because I tell all kinds of lies in a plausible and specious way, but also because everything in my story is a more or less comical parody of one or another of the poets, historians, and philosophers of old, who have written much that smacks of miracles and fables. . . . My lying is far more honest than theirs, for though I tell the truth in nothing else, I shall at least be truthful in saying that I am a liar. . . . Be it understood, then, that I am writing about things which I have neither seen nor had to do with nor learned from others—[things] which do not exist at all and . . . cannot exist. Therefore my readers should on no account believe in them.[12]

Lucian is toying with his reader from the outset. "Once upon a time," he continues, "setting out from the Pillars of Hercules and heading for the western ocean with a fair wind, I went a-voyaging" (I. 5, p. 253). What follows is a calculated swirl of events, adventures, monsters, armies, wars, escapes, and escapades. Sailing west, the narrator encounters an island of wine springs and is then struck by a violent whirlwind and carried aloft to a land that appears to be another island, "bright and round and shining with a great light," having "cities in it and rivers and seas and forests and mountains" (I. 10, p. 259). This he infers to be some unknown portion of the Earth but later discovers it is the Moon. Upon landing, he meets Endymion, who is at war with Phaethon and the Sun people. He is drawn into battle, captured, released, and finally entertained for a period of seven days before being allowed to leave the lunar surface. While there, he observes "strange and wonderful things": people born from the calves of men; clothing made of glass; eyes that are removable; noses that "run honey." He notes—in an apparent reference to Philolaus (see chapter 2)—that the beings there "are not subject to calls of nature" (I. 23, p. 277). Following this stay, he sails off again, this time to the farther reaches of space, where more adventures ensue. At the very end of the work, lost among new islands and wonders, Lucian's narrator sets sail one last time: "When the light of day began to show, we saw land and judged it to be the world opposite the one which we inhabit" (II. 47; p. 355). A second whirlwind attacks and destroys the ship, casts the survivors upon this "world opposite," whose stories, Lucian promises us in the final line, "I shall tell you in the succeeding books." (II. 47, p. 357). This world opposite, or opposite world, is indeed the Moon once again.

Lucian's narrative is a direct satire of *Wonders beyond Thule* and other early tales of similar marvels (e.g., Ctesias's *Indika* about India). The centrality of the Moon as a region of military conflict, attempted colonization, and war seems perfectly geared as a satire on Rome and its attempts to conquer lands "beyond the known world." But Plutarch is also a possible target here. One might recall that the final portion of *The Face in the Moon* offers us a deadpan, secondhand explorer's log of mythic adventures beyond the Pillars. The lunar surface as a somber, Neoplatonic region of half-cleansed souls is exactly the sort of depiction that Lucian's rhetoric of impossible witness would find worthy of parody. Whereas Plutarch's Sulla speaks of love between the Sun and Moon (this being the reason for their conjunctions), *A True Story* informs us that the Moon is at war with the Sun people because of attempts at colonization, acts that had aroused the greatest "repulsion" (I. 12, pp. 261–263).

Icaromenippus, *or the Sky Man*

Lucian twice used the lunar surface as a landscape for satire. A second Moon voyage, *Icaromenippus,* was launched for no less adventuresome reasons than those of *A True Story.* The hero of this tale, Menippus (named for the famous satirist), desiring to learn the secrets of the heavens, ties together the wing of a vulture with that of an eagle, fits them with two hand straps, and after a bit of practice soars skyward. He justifies his journey this way: "Above all, the peculiarities of the Moon seemed to me extraordinary and completely para-doxical, and I conjectured that her multiplicity of shapes had some hidden reason."[13] This reason Menippus had earlier sought from the philosophers— "those high thundering gentlemen"—who had written so copiously on the cosmos and its meaning. Finding a variety of theories and opinions, however, Menippus "despaired of hearing any truth about these matters on Earth and thought that the only way out of [his] dilemma would be to get wings somehow and go up to heaven" (10, p. 283).

Lucian has his hero perform a simple yet consequential deed that no other writer had yet attempted either through fiction or speculation. Upon reaching the Moon, Menippus rests for a while and turns his gaze downward upon the Earth, whereupon "the life of man in its entirety disclosed itself to me, and not only the nations and cities but the people themselves as clear as could be . . . all the life that the good green Earth supports" (12, p. 289). What does he then see?

> I saw Ptolemy lying with his sister, Lysimachus' son conspiring against his father, Seleucus' son Antiochus flirting surreptitiously with his stepmother, Alexander of Thessaly getting killed by his wife, Antigonus committing adultery with the wife of his son, and the son of Attalus pouring out the poison for him. . . . Similar things were to be seen going on in Libya and among the Thracians and Scythians in the palaces of kings. . . . [The] common people were far more ridiculous—Hermodorus the Epicurean perjuring himself for a thousand drachmas . . . the orator Clinias stealing a cup out of the Temple of Asclepius, and the Cynic Herophilus asleep in the brothel. . . . In brief, it was a motley and manifold spectacle. (15, pp. 293–295)

The Moon provides a purifying distance from which to view all of human sin, which, as Lucian teases, apparently composes the whole of human existence. *This* is the lunar secret, this distance that tears the mask off the close and familiar. The lunar disk becomes a lens focused on terrestrial realities, magnifying by miniaturizing, allowing a greater span of vision by physically reducing the Earth to a region "below."

There is much originality in this. Lucian has given the Moon a new moral dimension. Rather than a complex expression of divine order in heaven, the lunar orb is a revelation of chaos on Earth in more than one sense. In much of the remainder of the tale, the Moon herself convinces Menippus to petition Zeus to do away with the philosophers, and Zeus proclaims to the gods that this will, in fact, be done. The Moon's lament for her own image is instructive:

> I am tired at last, Menippus, of hearing quantities of dreadful abuse
> from the philosophers, who have nothing else to do but to bother
> about me, what I am, how big I am, and why I become semi-circular
> and crescent-shaped. Some of them say I am inhabited, others that I
> hang over the sea like a mirror, and others ascribe to me—oh, any-
> thing that each man's fancy prompts. Lately they even say that
> my very light is stolen and illegitimate, coming from the Sun up
> above. . . .
>
> But am I not aware of all the shameful, abominable deeds they
> do at night, they who by day are dour-visaged, resolute of eye, ma-
> jestic of mien, and the cynosure of the general public? Yet although
> I see all this, I keep quiet about it, for I do not think it decent to
> expose . . . their life behind the scenes. . . . But they for their part
> never desist from picking me to pieces. (20–21, pp. 303–305)

The Moon's plaint seems to be that the superabundance of opinion about her, a corpus of thought so often full of mutual exclusions and idle curiosities, is itself an expression of an overflow of human sin. Lucian's is not the standard Romantic line, "we murder to dissect" (Wordsworth). He instead illuminates disconnections between the daily lives of individuals and their presumed search for truth, between their desire for knowledge and their motive for self-aggrandizement as a means of gaining fame as "thinkers."

In the end, the image of the lunar body that Lucian gives us is a serious one. Such, of course, is the very object of satire. By using the lunar disk as a fun-house mirror for the inhabited Earth, the author reveals that the versions of the heavens proffered by Greek philosophers were based on ethical precepts—systems of universal order, harmony, and perfection that might serve as models for human thought and behavior. Lucian's Moon wants to be "left alone." "I cannot remain in my place," she says, "unless [Zeus] destroys the natural philosophers, muzzles the logicians, razes the Porch, burns down the Academy, and stops the lectures in the Walks": only then can "I get a rest" (21, p. 305). Only when philosophy is dead will the heavens be able to live in peace. There is this quality of anti-intellectualism in Lucian's work. He was certainly not above such sentiments, bearing as he did the marks of Roman sensibility. But this is because he has exploited the Moon, in somewhat traditional fashion, as a literary device to propose an ethical high ground. The philosophers cannot know the Moon properly, for they are themselves composed of "corrupt material." Their incapacity for the "good" renders them incapable of the "truth." Their very ideas stain the lunar face with terrestrial impurity. It is Lucian's view that only the ethically just individual can comprehend the substance and order of the universe. The true nature of the lunar orb will therefore remain hidden until the arrival of such an individual.

Ptolemy, Pliny, and Lunar Morality in the Roman Era

In the first few paragraphs of Book 1 of his masterwork, *Syntaxis mathematica,* Ptolemy makes distinct claims for the superior value of studying astronomy. Moral, aesthetic, and scientific dimensions are all merged into a sort of rhetorical compound: "Of all studies this one especially would prepare men to be perceptive of nobility both of action and of character: when the sameness, good order, proportion, and freedom from arrogance of divine things are being contemplated, this study makes those who follow it lovers of this divine beauty and instills—as it were, makes natural—the same condition in their soul."[14]

Such ideas of participation in the divine harmony through contemplation of its forms go back to a time before Plato. Yet by "astronomical knowledge," Ptolemy does not mean philosophical meditation inspired by a rudimentary perception of spherical shapes; he means mathematical astronomy in all its speculative glory, i.e., advanced geometry and trigonometry. It was this theoretical, abstract knowledge of the heavens that had commerce with the divine, not philosophy per se.

As expressed in two other works, *Tetrabiblos* (The Four Books, an astrological text) and *Planetary Hypothesis,* Ptolemy considered the Earth, including everything in the sublunary realm, to be corruptible and imperfect. Mathematics floated above this realm into a purifying domain that began just beyond the sphere of the Moon. Philosophy—for Ptolemy as for his near contemporary Lucian—was very much a part of the earthly world. Indeed, Ptolemy's inherited belief in the superior moral influence of astronomy was in harmony with his astrological presumptions, the heavens having many forces of influence over human conduct and destiny. In *Tetrabiblos,* the Moon is a mixture of pure and impure; it is solid yet moist because it is near to the Earth and thus draws upon the terrestrial vapors. It has some qualities of warmth "borrowed" from the Sun, but its influence on earthly bodies is to make them soft and subject to putrefaction. Recall Plutarch's comment: "[T]he Moon, if she has nothing corrupted or slimy in her . . . has got open regions of marvelous beauty and mountains flaming bright." In *Tetrabiblos,* as well as in *Almagest,* Ptolemy carries forward strands of thought and imagery that reach back to Plato and the pre-Socratics. The Moon appears in his work in different guises depending on the narrative context: as a geometric figure in discussions of planetary motion; as a reflecting sphere where the subject turns to celestial light; and as a solid or semisolid domain of wetness, femininity, and decay in astrological discourses.

Ptolemy was a Greek writing in the midst of a Roman world. His work reflects the heritage of Hellenistic scientific thought—with its mixture of theoretical sophistication and mysticism—not the selective appropriation of that heritage by Rome. Most of Ptolemy's works were apparently composed in Alexandria, Egypt; this is almost certainly true of *Almagest,* the culmination of scientific astronomy in antiquity. From whom, therefore, can we learn the views of the Moon that were held by Roman thinkers and writers? We might turn to Plutarch and Lucian for reference, but these men were also Greeks by birth and wrote their works in Greek.

The acknowledged authority on Roman science, or rather natural history, was Pliny the Elder (A.D. 23–79), whose *Naturalis historia* in thirty-seven books represents a vast, rambling encyclopedia covering nearly every major area of knowledge known to his countrymen.[15] Pliny's work includes a chatty survey of the heavens, partly inspired by Aratus (probably in the translation by Cicero). Within this survey, astronomy is largely reduced to a series of descriptions and names, often with astrological inklings. Regarding the Moon, the author has a number of specific things to say. After calling the lunar body "our teacher as to all the facts that it has been possible to observe in the heavens," he goes on to list these:

> (1) that the year is to be divided into twelve monthly spaces . . . ; (2) that she [the Moon] is governed by the Sun's radiance as are the rest of the stars . . . ; (3) that consequently she only causes water to evaporate with a rather gentle and imperfect force . . . ; (4) also that . . . she shows as much light from herself to the Earth as she herself conceives from the Sun; (5) she is indeed invisible when in conjunction with the Sun . . . ; (6) but that the stars are undoubtedly nourished by the moisture of the Earth, since *she is sometimes seen spotted in half her orb, clearly because she has not yet sufficient strength to go on drinking—her spots being merely dirt from the Earth taken up with the moisture.* (II. vi., p. 197; italics added)

Pliny employs the term "maculas" for the lunar maria *(maculas enim non aliud esse quam terrae raptas cum humore sordes)*, a term that had been in general usage for centuries and whose notion of "spottedness" connects directly with the moral-philosophical tradition of viewing the Moon as tainted by corrupt, imperfect qualities. Pliny, however, gives this a wholly concrete sense, saying that the lunar surface is not merely composed of lower matter but is actually full of mud *(sordes)!*

CONCLUSION

Ptolemy, Lucian, Plutarch, and Pliny: each of these authors represented an endpoint. Whereas Ptolemy brought to an apex the tradition of mathematical astronomy, then more than five hundred years old, Lucian carried the still older rhetorical uses of the Moon to a final level, where earlier conventions of speculation and travel writing were turned back on themselves as satire. Plutarch took the philosophical dialogue to a manneristic extreme, weaving in many styles of discourse, many areas of knowledge, dozens of literary references, a host of speakers, and ending, not with a concluding statement of presumed fact, but with a nostalgic myth that looks back across four centuries to Plato, true father of the form. Finally, Pliny stitches together a Moon that is solid and reflects the Sun's light, a concept from astronomy proper, with more ancient precepts about its corrupt nature. All of these authors present an image of the lunar surface that is linked directly to earthly matter and, therefore, to a mixture between pure and impure substance. To the degree that the Moon is derived from the Earth, it shares the Earth's own material, moral, and spiritual difficulties.

At the edges of antiquity, the Moon remained a phenomenon of many faces. What happened thereafter was a parting of the ways. In Europe, the likes of Aratus and Pliny and the question of a plurality of worlds would hold sway. Toward the east, Islam would take up the Greek intellectual corpus, including Plutarch, Aristotle, and Ptolemy, and submit it to extensive commentary and improvement. Both of these paths met again, in the twelfth century and later years, when the full heritage of Islamic/Greek science was translated into Latin. It is to this portion of the Moon's history on Earth that we now turn.

4

The Moon and Medieval Science

TEXT AND IMAGES BEFORE THE
TWELFTH CENTURY

THE MOON AND THE SEVEN LIBERAL ARTS

*I*n the last years of the Roman Empire, and thus at the brink of the Middle Ages in Europe, several works involving astronomical subjects were written that set the course of knowledge in this area for eight hundred years. Ptolemy's writings were not among them, nor were those of Hipparchus, Archimedes, or Plutarch. Nearly the whole of Hellenistic physical science slipped beyond the reach of what is commonly called "the West." The Romans had found little use for its astronomy, with its difficult, mathematically demanding thought. Their taste was for a more simplified, descriptive portrait of the heavens that could be used in weather prediction, calendar making, and astrology. The texts of Greek science journeyed eastward, first to Byzantium, and then, under pressure of persecution, to Syria in the caring hands of Nestorian and Monophysite Christians, finally to rest in the appreciative libraries of Islam.[1]

Early medieval astronomy among the Latins, meanwhile, was derived from several major texts. These works mainly dealt with planetary movements or with more general celestial (stellar and planetary) phenomena. The first group included three texts in particular: *Commentary on the Dream of Scipio* by Macrobius (fourth century A.D.), *Commentary on Plato's Timaeus* by Calcidius (fourth century A.D.), and most importantly, *The Marriage of Philology and Mercury* by Martianus Capella (late fifth century A.D.). Celestial astronomy was represented by Pliny and Aratus, whose works were discussed in earlier chapters.[2] If we think in terms of the "liberal arts" and the importance they acquired since the Carolingian era, the most influential work would have to be the brief encyclopedic handbook, *The Marriage of Philology and Mercury*. This book was written in a mixture of styles only a decade or so after the final collapse of Rome, a context that adds a touch of irony to the literary qualities and intent of this work, proposing as it does a flowery celebration of knowledge and its assumption to the heights of heaven.[3] Writing at the threshold of the Dark Ages, Martianus holds forth on the seven *artes*—grammar, dialectic, rhetoric, geometry, arithmetic, harmony (music), and astronomy—as if the gates to final wisdom and the path to divinity had finally opened.

The Marriage of Philology and Mercury is divided into two basic parts. There is an allegorical prologue announcing the wedding of the god Mercury (eloquence) to the maiden Philologia (learning), whose seven slave attendants introduce themselves in the second part through speeches that summarize their respective "art." In contrast to the exaggerated, baroque language of the first two books announcing the marriage, each of the following speeches is a concise, straightforward epitome of its subject. In the case of astronomy (book VIII), for example, scholars agree that Martianus's treatment represents "the most orderly and comprehensive treatment of the subject by any Latin manuscript extant."[4] It presents a discussion of the planets and fixed stars, the ecliptic, constellations, and eclipses, focusing above all upon celestial motion. This is obviously the legacy of Greek mathematical astronomy reduced to textual description. Martianus, however, has not a word to say about the substance or nature of any of the planets, only their geometric movements.

In the allegorical prologue, another type of astronomy can be found:

Then the bearers picked up [Philology's] palanquin and with great effort carried her aloft. Borne up by their buoyancy they rose

126,000 stades, and completed the first of the celestial tonal intervals; then the maiden entered the circle of the Moon, and in those vapors suitable to a goddess . . . she saw a soft spherical body composed with the smoothness of dew from heaven, reflecting, like a gleaming mirror, the rays of light that fell upon it. In it there appeared the sistra of Egypt, the lamp of Eleusis, Diana's bow, and the tambours of Cybele. Changing in color, and threefold in form during its cycle, it shone with awesome majesty. Although it was thought to be horned and rough, yet when it emptied itself it showed, according to its season, a cat or a stag, or any of four appearances.[5]

In a few short sentences, Martianus evokes an entire tradition of lunar imagery lying outside the writings of poets, astronomers, and popularizers and mixes it with one of the most ancient conventions of all (femininity, moistness). In the face of the Moon itself, he tells us, there are several appearances that resemble animals, objects, even musical instruments. The sistra (a curved metal rattle used at Egyptian festivals), a lamp or bow, the tambour (a type of drum), a cat, and a stag can all be seen in the patterns of light and shade on the lunar surface. Such visions hail from prominent mystery cults in Egypt, Babylonia, Greece, and Rome,[6] in particular those associated with Isis, Cybele, Ceres, and Diana, and had probably evolved into popular folklore by the time Martianus wrote of them.

Until now, the discussion has not included these types of lunar images because they were not mentioned by the Greek philosophers or astronomers, including Plutarch, at least not in the writings that survive. These images reveal a high degree of close observation of the Moon's surface appearance. The figure in the Moon was given other aliases during antiquity and the early Middle Ages. Folklore records not only the cat and stag but also a rabbit, a reclining woman or girl, and a pair of young children with a pole lodged between them.[7] Astrologers claimed the ability to divine future events, particularly those concerning weather, pregnancy, and mental condition, by noting the exact positions of dark and light on the lunar surface (positions, of course, that never changed). "Scrying" the Moon, as it was called, even included reading earthly happenings by the figure that appeared in a reflection of the lunar face on a pond, stream, or lake.

None of these images or aspects of the lunar imagination, as visual as they are, appear to have been recorded in the artistic works that remain from late antiquity or the medieval period. One reason for this may have been the unwillingness to portray a body ripe with omen and magic for fear of ill effect or reprisal. The question, however, still remains: What types of drawings of the Moon *did* exist during the Middle Ages, and what do they indicate about the history of observation and the development of scientific illustration?

FIGURES DRAWN: THE PAINTINGS AT DURA AND EARLY SYRIAC GOSPELS

Some of the very earliest painted images of the Moon that remain in existence, apart from allegorical works depicting Selene and Luna, appear in illustrated literary texts of the fifth century A.D. Possibly the oldest of these manuscripts is the *Vergilius Vaticanus,* an illuminated version of the Aeneid that includes a scene of the sack of Troy, showing in brilliant color the city walls, Trojan Horse, attacking Greeks, and in the background, a ship pulled up on the shore beneath a crescent Moon and several gilded stars.[8] Inclusion of the Moon appears to have been done simply for the sake of background and suggests that such use was common or at least well established.

Within Jewish and early Christian society, on the other hand, the lunar orb took on iconic uses, with the oldest known images appearing in sixth-century religious art of the Near East. Perhaps the most striking example is included among the magnificent wall paintings at the Dura synagogue in southern Israel, preserved by a particular, if tragic, set of circumstances. Dura was a small Roman military outpost along the eastern margins of the empire. About A.D. 256, this largely Jewish settlement was attacked and destroyed by an advancing Persian army. In a desperate attempt to stave off the inevitable, the inhabitants shored up the weakest wall of the town by filling with earth the buildings lined alongside it, including the synagogue with its newly painted interior. The attempt proved futile; Dura was left a buried and unknown ruin for more than fifteen hundred years. Not until the twentieth century was the town excavated and the beautifully preserved wall paintings brought to light once more, at which point they became the subject of intense and careful study, with profound results for the history of Jewish art, literature, and culture.[9]

Figure 4.1. Sketch of the upper part of a wall painting at Dura synagogue, dated about A.D. 256, showing the head of a man (Moses?) and above it the Sun, seven stars (Pleiades?), and a crescent Moon, all in stylized fashion. Redrawn by Floyd Bardsley.

Painted on one of the wing panels are four portraits of unknown figures, possibly Moses (the upper two figures), Ezra, and Abraham.[10] These are the only portraits in the entire series of paintings and are thought to depict important figures. One portrait shows a man with white hair and beard wearing a Greek robe; above his head stretches an arc of the heavens containing the Sun on the left, seven stars, and the crescent Moon on the right. A rendering of the upper part of this image is shown in figure 4.1 and should be compared with the clay tablet shown in figure 2.1.

The similarities in the depiction of the stars and crescent Moon are striking. It is as if a thousand years had yielded no advances in portrayals of the heavens. The artists at Dura, it seems, employed an image pattern that was traditional, well established, and easily understood by the populace at large. To say even this much is speculative; yet there is at least the implication that the basic arrangement and style of figures 2.1 and 4.1 represent a pattern that had remained stable for centuries and had been copied on numerous occasions. This is also suggested by certain sculptural works that offer astrological predictions and include identically pointed stars, a crescent Moon, and one or more allegorical versions of the Sun and planets.[11]

The image in figure 4.1 in particular recalls an important and much used commentary on the Old Testament, by the Hellenistic Jewish scholar Philo Judaeus (ca. 20 B.C.–A.D. 50), in which the following lines occur:

> He [Moses] gathered together a divine company, that is the elements of the universe and the most effective parts of the cosmos, namely the earth and heaven. . . . In the middle between these he composed hymns using every musical mode and every type of interval in order that men and ministering angels might hear. . . . The angels would also be strengthened in their faith if a man clothed in his mortal body could have a power of song like the Sun, Moon, and the sacred choir of other stars, and could attune his soul to the divine instrument, namely the heavens and the whole cosmos.[12]

Moses, it appears, is at the end of his life, about to enter the ranks of the ethereal. He sings the "perfect song while yet in his body" and (in lines that follow the above passage) expresses his love and concern for his people, scolds them for their sins, and gives them hope for their future. His voice is "attuned" to the "music of the spheres," filled with the force of heaven. He sings the Platonic harmony of harmonies that will bear to the Israelites a depth of religious faith and classical virtue worthy of the angels.

When Moses dies, the Old Testament tells us, another great leader steps forward, the warrior Joshua. The Lord sends Joshua to lead the armies of Israel across the River Jordan and there establish a final homeland. In the midst of a great battle in the valley of Ajalon, the Sun begins to set while victory remains undecided. As he sees the land darkening, Joshua calls out to God for aid:

> "O Sun, stand thou still upon Gibeon;
> And thou Moon, in the valley of Ajalon."
> And the Sun stood still,
> And the Moon stayed,
> Until the people had avenged themselves
> Upon their enemies.
> —Joshua 10:12–13

There is no small poignancy in thinking that such words may have inspired the artist at Dura, perhaps only days before the Persian army descended upon the town. At the same time, there are difficulties with this interpretation: there are no stars mentioned in the account of Ajalon, and Joshua was not a man whitened by age but was vigorous and in his prime. Could it be that the artist wished to combine more than one invocational theme in his image? The question must remain unanswered. What is important is that an ancient model of the heavens was borrowed for a very specific religious purpose, one that may well have altered its passage into the future.

This is suggested by two other early works in which the Moon appears in painted form. One of these is among the oldest known manuscripts in Syriac, the Rabbula Gospels, dated A.D. 586, so named because they were written and illuminated by the bishop Rabbula. The codex contains richly colored illuminations said to be derived from earlier wall paintings or mosaics in Palestinian and Syrian churches. Among the most striking of Rabbula's images is a full-page depiction of the Crucifixion and Resurrection. In this miniature, over each arm of the cross and above a valley between two towering and crudely drawn mountains, stand the Sun and Moon, this time with the Moon on the left in crescent phase, resting in the center of a dark disk. A Syriac Bible of the next century contains the portrayal previously discussed as a possible interpretation for the painting at Dura: Joshua at Gibeon, painted in full battle dress, sword in hand, pointing upward to the Sun that shines down on the left while the crescent Moon hangs on the right.[13] The body of Joshua has a slight Hellenistic sway to it as well as skillful shading, once again implying the use of existing models. Unlike the Rabbula Crucifixion, there is little complexity in this image. It is merely a literal rendering of the story about Joshua at Ajalon, suggesting that use of the Sun and Moon as iconographic elements was common practice by this time.

ICONOGRAPHY OF THE MOON IN
EARLY CHRISTIAN ART

Associating the Sun and Moon with the dying Christ defines an artistic motif that continued unbroken throughout the entire medieval period, during the Renaissance, and even into the sixteenth century. Images of the solar disk and crescent Moon above the arms of the cross in representations of the Crucifixion can be found in hundreds of works, not only in illuminated books, but also in ivory carvings, chasubles, reliquaries, bishops' robes, tabernacles, frescoes, and paintings; i.e., in nearly every genre of Christian art. Inherited from antiquity, the time of the tablet shown in figure 2.1, this motif constitutes one of the oldest iconic threads within the greater weave of Christian art.

The meaning of this association is not clear. Some evidence suggests that the Moon, because of its link with corrupted matter, inconstancy, and the terrestrial sphere, was present in images of the Crucifixion to symbolize the mortal, earthly side of Christ's nature. Such an interpretation is implied by traditional connections between the Virgin Mary and the Moon, based on commentaries concerning the vivid and dramatic imagery in *Revelation* 12:1–4, which reads, "A great and wondrous sign appeared in heaven: a woman clothed with the Sun, the Moon beneath her feet, and on her head a crown of twelve stars. She was about to give birth and in the agony of her labor cried out. Then a second sign appeared in heaven: a huge red dragon with seven heads and ten horns and seven crowns on the heads. With his tail he swept a third of the stars out of the sky and flung them to the Earth."

From as early as St. Irenaeus (second and third century A.D.), commentators spoke of the woman either as the Virgin Mary or the Holy Mother Church, with the other portions of the vision attracting a wider array of associations (e.g., the dragon as Satan, a host of false churches or unbelievers, or the Roman Empire). An especially influential commentary, produced in the sixth century A.D. by Pope Gregory the Great, favored the choice of Mary as the symbolized presence and said that the Moon lay crushed beneath her feet, representing "all fallen, mutable, and earthly things."[14] Taken together, the Sun and Moon in images of the Crucifixion are therefore likely to encompass good and evil, divine and fallen, sacred and profane, immortal and mortal.

The Crucifixion, though the main setting for nonscientific images of the lunar orb in medieval Europe, was not the only one. Alongside the "sullied" notion of Pope Gregory the Great, there was a contrasting, favorable association between the Virgin Mary and the Moon, revealed by the inclusion of the lunar crescent in some paintings of the Annunciation and the Coronation of Mary. Such associations appear to be much more typical of the later Middle Ages, from the thirteenth century onward, and may well be related to the introduction of Aristotelian ideas and the influential commentaries on them by authors such as Averroes. As interpreted by the Arabs, Aristotle believed in a crystalline lunar sphere, set aglow by the Sun's light and made of a single, homogeneous substance whose dark and light patches were caused by denser or more rarefied conditions. The suggestive purity of this interpretation, coupled perhaps with the ancient connection between the Moon and notions of womanhood, helped further the association with the Virgin Mary, a link that became especially common and debated in the later Renaissance and early seventeenth century.[15]

These connections strongly suggest that the oldest tradition of the Moon as painted image came not from science or philosophy but from literature and religion. These were the realms, after all, in which the lunar body had long played an important role and had drawn to itself the greatest spectrum of interpretive imagery. It is no surprise that this would remain the case for some time.

CAROLINGIAN ASTRONOMY: TEXTS AND AESTHETICS

Beginning with the Carolingian revival of classical learning in the eighth century, medieval authors interested in the heavens regularly turned to Martianus Capella, Pliny, and Aratus. They also consulted the works of Cassiodorus, Isadore of Seville, and most importantly the Venerable Bede, whose *De rerum natura* (On the nature of things, ca. A.D. 703) and *De temporum ratione* (On time reckoning, A.D. 725) laid out the celestial hierarchy in Christianized detail and injected a high level of arithmetic calculation into uses of the heavens for calendrical science. As reflected in the number of surviving manuscripts, however, the Carolingians preferred Martianus's *The Marriage of Philology and Mercury* (perhaps the primary textbook of the entire succeeding medieval period), Pliny in a unique form, and illustrated versions of Aratus. Pliny gained influence through a collection of excerpts from *Naturalis historia* concerned with the planetary orbits and motions. These passages were selected, partially rewritten, and assembled about A.D. 809 for use as a teaching aid and a brief, theoretical reference.[16] *Phænomena* of Aratus (both in Cicero's translation and in that of Germanicus Caesar) was resurrected in beautifully illuminated manuscripts, of which there are a fair number remaining today, some bearing among the most striking artistic images painted during the entire early medieval period.

The Carolingian revival marked a strong return to Roman learning in particular (due to the constraints of Latin) and among the sciences, to astronomy above all. A pronounced effort of the late eighth and ninth centuries was to recover and solidify such learning by building "workshops of knowledge" (a phrase identified with Charlemagne) in the form of cathedral schools that would then serve as essential sources of strength for the Church and a new Holy Roman Empire by becoming centers of literary and artistic endeavor, where manuscript copying, collecting, and study especially would occur. This emphasis on the power of words and language—a hallmark of Carolingian achievement[17]—helped enforce a hierarchy of image making, whereby drawn and painted images were viewed as secondary to writing and reading. Inspired by the thought of St. Augustine, St. Gregory, and others, *Libri Carolini* (Caroline Books) expressed a central tenet of the times in stating that words and images embodied knowledge in different ways—words by their direct and "undefiled" transfer of truth, images by their ability to act as narrators and teachers to the illiterate and by their powers for conveying beauty, skill, and revelations of divine order. As one recent scholar has commented, "It is within the wary acceptance of the didactic value of pictures and the conviction that writing was more reliable, more truthful and unambiguous, that Carolingian book painting has to be seen."[18] This is true not only of book painting, however, but of all forms of scientific illustration.

Within this framework, the idea of "observation" was suspended, neither lost nor liberated, but left on the wayside. Naturalistic portrayal of the physical universe appeared only where classical models were closely followed.

ASTRONOMICAL ILLUSTRATIONS: MAJOR TRENDS

Prior to the middle of the eighth century, the major astronomical work to regularly bear diagrams or illustrations was the *computus*. This type of text had the primary function of using planetary cycles to calculate the specific dates of important holy days (especially Easter) and the seasons. Bede, drawing on Pliny and other authors, had set the course for this genre in the middle seventh century, expanding it to embrace the arithmetic of the heavens as a whole, including the zodiac and planets.[19] Before Bede, the computus was a bare-bones document, meant to serve purely practical uses without diagrams.

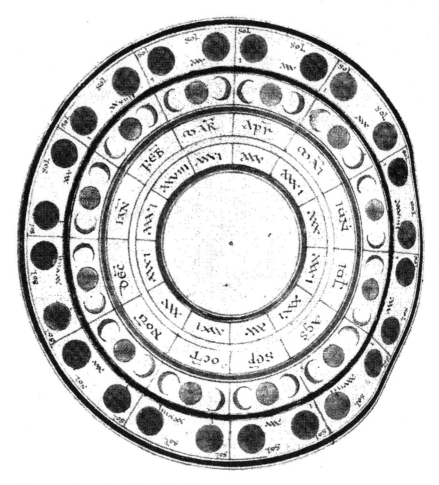

Figure 4.2. Images from the Hildebald manuscript, dated between A.D. *798 and 805 and archived in the cathedral library in Cologne, Germany (Cod. 83). The Moon is depicted in various forms, including classical personification (the goddess Luna), circular orbits, and phase diagrams.*

From Bede onward, however, new types of imagery were employed: circular charts, diagrams showing the arrangement of the celestial bodies, drawings depicting the orbits of the planets and the phases of the Moon, and actual illuminated paintings of the constellations and planets in personified form (fig. 4.2). This tendency to include illustrations expanded considerably in the late eighth and ninth centuries, when it became common practice to

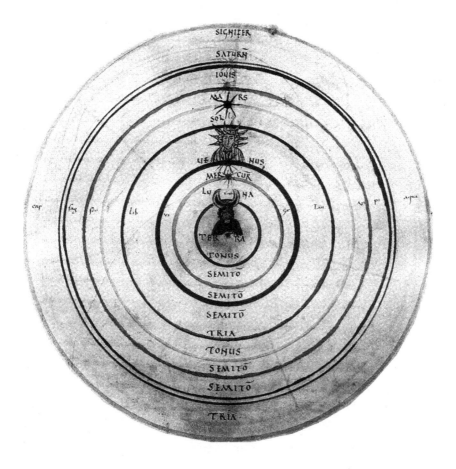

produce sourcebook-type compilations, i.e., astronomical manuscripts made up of selected passages from a wide range of different authors.

One such text, known as the Hildebald, is a multibook computus displaying a magnificent abundance of diagrams, apparently collected from several versions of Bede's *De temporum ratione* (On the logic of time) and other books.[20] The Hildebald manuscript was produced between A.D. 798 and 805 and was quite likely used by scholars in Charlemagne's court at Aachen. By almost any standard, it is a remarkable and visually magnificent work. It provides nothing less than a grand combination of existing types of astronomical illustration: geometric diagrams, Sun/Moon orbit and phase cycles, intricate illustrations of planetary apsides and latitudes, drawings of the five terrestrial zones, artistic paintings of the constellations, and more.[21]

As noted with regard to Martianus and the Moon, we find in the Hildebald computus more than one species of "astronomy." There is an astronomy of arithmetic calculation, of decorative order and pure geometry, and of myth. Such diversity acts to contradict any unified vision of the universe, any single "view." Each type of image, of course, bears its own internal logic and cognitive demand; it calls upon the reader in a particular way. Arithmetic diagrams and tables posed the universe as ruled by numbers; geometric illustrations offered it as a representation of engineered order and fixed spatial relations, reminiscent of gears in a mechanical device; allegorical figures of the planets proposed that one see the skies in classical aesthetic terms, with the entire Carolingian effort of *renovatio*. The bringing together of all of these "astronomies" reflects the eager and often uncritical revival aspect of the Carolingian intellectual movement.

How does the Moon appear as an image in the Hildebald manuscript? Figure 4.2 provides a sampling that corresponds to each of the different astronomies mentioned above. We see the Moon reduced to the circularity of its orbit, expanded into its succession of phases, and rendered into the form of a goddess. All of these types of images existed in antiquity. What is new here is their inclusion within a single work. Carolingian astronomy, in other words, could be a kind of collector's cabinet. Its contents, however, were more limited than those of antiquity: it shows no interest in speaking of the actual appearance of the lunar orb or its composition. The "spottedness" of the lunar surface was not a subject of discussion beyond the continual rehearsal of Pliny's characterization that this constituted "merely dirt from the earth taken up with the moisture." It is ironic that the most naturalistic element offered by any of the images found in the Hildebald treatise—that of earthshine (light reflected from the Earth to the Moon, faintly illuminating the lunar disk during the crescent or gibbous phase)—appears in the painting of the mythic figure Luna.

PLINIAN DIAGRAMS AND THE
TRADITION OF MATHEMATICAL HEAVENS

Plinian geometric illustrations carried forward the simpler traces of the tradition of Hellenistic mathematical astronomy. Again, this was a tradition that was not much interested in observing, studying, and depicting the lunar surface but instead in describing celestial movement. Even questions about why the planets change size and color, for example, were answered within this tradition through variations in latitude and the angle of the zodiacal circle. The concept of Venus or the Moon, as "other worlds" in the physical sense, does not attain a following at this time.

At first blush, the range of diagrams in the Plinian handbooks seems considerable. There are various types of concentric drawings depicting planetary order and the harmonic intervals between orbits. There are complex circular illustrations showing planetary apsides and latitudes, with up to thirteen concentric circles representing the interval degrees of the zodiac, over which are imposed the different eccentric orbits for the planets and the Sun. Such diagrams are difficult to read; at their worst, they become manneristic and suggest that pedagogy has been left behind in favor of a certain infatuation with visual relationships on the page, with the sheer power of drawing itself. At this point, such figures no longer illustrate the order of the solar system. Instead they express the impossibility for any single diagram in this tradition to include all planetary motions, and therefore the inevitability that such a diagram would eventually move from the realm of "science" (knowledge) into that of aestheticism. As early as the later ninth century, this seems to have been realized because new rectangular drawings were invented to isolate and graph more simply the position of each planet through time.[22]

The mathematical cosmos had gained a number of forms by this time, both in image and in text. This development directly linked medieval Europe to antiquity by showing that the real work of astronomy was to comprehend the planets and stars through their movements, positions, and cycles. The absence of the grand Ptolemaic conceptual scheme in early medieval Europe did not especially matter in this regard. Such absence prior to the twelfth century has been often lamented as a sign of the low level of astronomy at this time. The heavens, however, were densely mathematical in a variety of ways. The skies were embodied in arithmetic calculations and the wholesale concern with time reckoning, which was primarily fixed upon the Moon and was central to European Christian culture from Gregory of Tours (sixth century) onward. On the textual side, too, a mathematical view prevailed. This is apparent in Bruce Eastwood's characterization of early medieval astronomy as a "series of discrete definitions of terms which emphasized observation [especially the fixed methods of observing and recording planetary motions], identification, and occasional tracking of celestial bodies."[23]

ARATUS IN THE MIDDLE AGES:
LITERARY ASTRONOMY RECLAIMED

It is difficult today to think of the beautifully painted miniatures adorning Carolingian versions of Aratus's *Phænomena* as "science." They instead appear to be the precious remnants of a once magnificent "art," that of book illumination. Emerging from an ancient song about the wheel of the heavens, these figures are among the most powerful and convincing testaments to the urge for *renovatio* during this period. As one scholar has written, they "might have stepped out of a Pompeian mural . . . to transmit to posterity the genuine effigies of the pagan gods and heroes that had lent their names to the celestial bodies."[24]

Of the various *Aratea* that have survived from this time (doubtless a small sampling of a much larger corpus), two in particular stand out as unique masterpieces. One of these is the Leiden *Aratus (Codex Vossianus Latinus 79)* archived in the university library at Leiden, dating perhaps from the early ninth century.[25] The miniatures that decorate this manuscript appear to be copies of much older models; this can be seen in the nakedness of many of the human figures used to depict constellations, their sophisticated balance and proportion, and their mixture of realism and stylization (fig. 4.3). These figures are solid, wonderfully shaded and colored, and full of exquisite detail, with the stars portrayed as small gold spots (many in the shape of diamonds) that do not in the least detract from the allegorical image. Although the purported center of interest, these stars are not much more than an afterthought: little concern has been given to the accuracy of their position and none at all to their relative brightness. True to Aratus, this is astronomy in the service of art, a literary astronomy illustrated in a manner loyal to the poetic text from which it stemmed.

Any hope of finding in such illustrated manuscripts a drawing of the Moon as a planetary body is therefore in vain. The closest one comes is in the final image of the Leiden *Aratea,* which is a map of the solar system, apparently the oldest known to exist, with the planets arranged as they would have

Figure 4.3. Images of the constellations from the Leiden Aratea, *dated eighth–ninth century. The figures include (clockwise from the upper left) Cygnus, Aquarius, Capricorn, and Sagittarius.*

appeared on March 28, A.D. 579.[26] Figure 4.4 reveals that the complexity of this diagram vies with that of any of the Plinian drawings noted above. It depicts the planetary orbits not as perfectly centered upon the Earth but as eccentric in shape, with Mercury and Venus actually orbiting the Sun— the solar system as depicted by Martianus Capella. Each planet, zodiacal constellation, and month receives its allegorical figure, mostly in the form of medallion-like images, with Luna shown in her chariot drawn by two oxen. To the unwary observer, Mercury, Venus, and the Moon might seem to

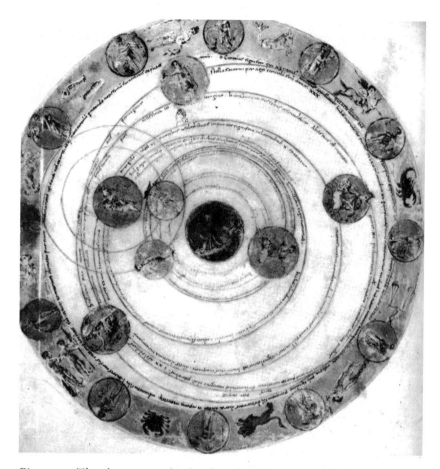

Figure 4.4. The solar system and zodiac from the Leiden Aratea. *The figure depicts geometric orbits with medallions holding allegorical figures for each planet. The Earth sits in the center; the Moon appears on its immediate right; the Sun lies to its left, circled by Mercury and Venus.*

occupy each other's orbits. Moreover, the months are placed in the wrong order with respect to the signs of the zodiac (one runs clockwise, the other counterclockwise). There appears to have been too much information packed into this one figure for the artist (or copyist) to keep accurate track of everything.

There is one other aspect to this early map of the solar system that requires mentioning because it brings us to the very heart of the question regarding scientific illustration. Pliny, as it happens, is present in this map in more than just echo. Around the circle of each planetary orbit are quotations

from *Naturalis historia* regarding perigee, apogee, and astrological exulta-
tion. The diagram was not considered finished or complete without significant use of text. This is because, unlike the miniatures, it was intended as a
pedagogic aid for learning the fixed and prescribed order of the heavens, not
their mythology. For Carolingian scholarship, images were not sufficient in
themselves to convey the actual science of astronomy.

The ultimate example here is a second Aratus masterpiece from the
ninth century, this one apparently written in France and then transported to
England, where it became a model for similar versions down through the
twelfth century.[27] In this illustrated manuscript, the body of each constellation is literally made up of words, a selection of passages from the *Astronomica*
of Hyginus, a second-century A.D. Roman author. Under each image is given
the relevant passage from Aratus's poem in the translation by Cicero. The extracts from Hyginus treat the number of stars, their names, relative brightness, and general position. They are written, moreover, in an antique script
known as *capitalis rustica,* a deliberate call upon Roman-derived, antiquarian
calligraphy. This unique method of portrayal offers the "authentic" science of
Hyginus (such as it is), by showing the bodily substance of each constellation, captioned by the "literature" of Aratus/Cicero. As with other Aratus
versions, very little attention has been given to the drawings of the stars
themselves. They are simply sprinkled here and there like so many grains of
gold, but their overall arrangement is just careful enough to suggest to the
celestial *illiterati* the figures of their respective namesakes.

THE MEDIEVAL MOON AND SCIENTIFIC ILLUSTRATION

At this early stage in the history of European science and art, text and image
are not yet separable as modes of expression. Browsing through whatever
herbals, bestiaries, lapidaries, or other technical manuscripts we might find
from this period, we see drawings of great beauty invaded by words at nearly
every turn. Plants grow and twine upward out of their own written names;
beasts of the forest or jungle pose like statues upon the stage of their description; medical guides unravel every nerve or artery with a caption or poetic
couplet; the universe spins and the planets hang upon wirelike circles and ec-
centrics over which revolve letters, numbers, names, and dates. Illustrations

are not trusted to do things on their own, nor were they for a long time to come.

The privileges granted to writing and the antiquarian impulse in Carolingian thought ensured that the Moon, drenched as it was in classical myth, would be portrayed most often in allegorical fashion. This took place well outside the precincts of technical expression and resulted in fascinating combinations of Christian and pagan imagery. In Bibles prepared and illustrated at this time, for example, personifications of Sol and Luna are fairly common. A particularly striking example is an ivory carving of the Crucifixion, executed about A.D. 870 and later set into a book of Pericopes for Henry II of England (early eleventh century). The image shows the Sun and Moon not as disk and crescent or as orbs borne by angels but as pagan gods in their respective wagons, being drawn by rearing steeds (Sun) or stolid oxen (Moon).[28] If such artworks were instrumental in reviving pagan allegory and symbolism about the heavens, no less was this true for literature. Images of the Moon as an emblem of love, past life, change, and inconstancy, fixed in written expression from the beginning, again became current: "As the Moon grows in light to full circle, yet soon vanishes in ever changing movements, so too the kingdoms of men grow and pass away."[29]

As an image for the imagination, the Moon remained richly variable. It could be associated with a pagan goddess, the Virgin Mary, the profane on Earth, the fleeting qualities of pride and human life, and the science of the skies. A metaphor as well as an allegory, the lunar body drew to itself ingenuities of portrayal that crossed the borders of many disciplines. This would not change, but it would undergo important alteration, particularly in the wake of Arabic science and its pivotal influence upon Europe.

5

The Later Middle Ages

FROM SYMBOLISM TO NATURALISM

CONTINUITY AND CHANGE IN LUNAR IMAGERY

The late Middle Ages mark a crucial period in the history of the Moon on Earth. New versions of the lunar orb and new discussions of its reality emerged in European thought and came to exist alongside those of the past. The twelfth to fourteenth centuries define the time when the physical makeup of the lunar body, including the "spots" on its surface, first came into sustained intellectual view. This took place at a time when symbolic predispositions toward material reality began to give way to, or be deeply modified by, a true naturalism based upon an empirical appreciation of the beauty, form, and concreteness of earthly phenomena. This overall trend would not reach its culmination until the early fifteenth century, most impressively in the art of Jan van Eyck. But its existence before that time is clear, and its outlook constituted a truly profound change over previous centuries.

Within astronomy itself, the main genres of illustration evident at the close of the Carolingian *renovatio* continued largely without interruption. Yet from the late twelfth century onward, with the full introduction of Greek and Muslim learning through translations from the Arabic, the balance between a "scientific" and a "literary" astronomy, between Plinian-type diagrams and allegorical pictures, came to an end. Greek-Muslim science (for this is what it was) effectively meant the demise of a literary-allegorical astronomy. At the same time, it brought with it a new division of the heavens into mathematical-theoretical and physical-philosophical halves, exemplified by Ptolemy on the one hand and Aristotle on the other. Once these authors were established within the new university system and adopted by schools and authors elsewhere, there was little room within astronomy proper for the high artistry of the Leiden *Aratea*. Illustrated versions of Aratus's poem, Martianus Capella's textbook, and the Plinian extracts continued to be produced, even after the introduction of printing. Yet serious students of astronomy had long abandoned such works by this time (or else used them sparingly) in favor of new texts that dealt more directly with geometric theories and the actual substance of the celestial bodies.

This general trend was itself aided by larger developments in medieval aesthetics. From the twelfth century onward, a new outlook on the natural world gained strength, inspired by a complex mix of Aristotelian ideas and gothic theology. Greek-Muslim thought helped urge influential writers such as Hugh of St. Victor to adopt the classical notion of the visual world as a source of true knowledge and to argue for the spiritual benefits received through appreciation of the realities of God's creation. Such notions, writes David Summers, "gave the deepest possible justification for an art appealing first of all to sense, because on such a view it is possible to ascend from the pleasing qualities of objects to the real presence of divine grace."[1] Art thus became a means to express divine harmony and order in its material manifestations. The centrality of art in European society grew, even as it looked outward to the nonhuman universe of forms.

The move toward more realistic portrayal of natural phenomena affected a wide array of expressions, from the ornamental borders of gothic manuscripts to the sculptures adorning the great cathedrals. The focus was clearly on organic nature first, on plants, animals, and humans most of all. Although by the fourteenth century this focus had widened to include many phenomena of the physical universe, it did not yet mean the complete end of literary-allegorical representations. The late medieval Moon retained something of its ancient iconographical substance, at least in part and for a time.

THE MAN IN THE MOON APPEARS

During the several centuries after Charlemagne's death, lunar depictions remained similar to those already discussed but with some important modifications. In astronomical treatises and in illustrations of biblical or literary texts, the goddess Luna began to shrink, indeed to change gender. From her full-scale form as a standing figure, perhaps borne by an ox-drawn wagon, she was scaled down to a head medallion, especially in Plinian-type diagrams, or to a man-in-the-Moon face with distinct male characteristics. The latter development appears to have been related to the increased amount of Crucifixion imagery, which, as noted previously, was commonly adorned with the Sun and Moon (each as a face within a disk) above the arms of the cross. By the twelfth and thirteenth centuries, this man-in-the-Moon image had become a standard icon in a wide range of new media, including woodblock prints, stained glass, fresco paintings, and more.[2] As before, the crescent phase was preferred, with the face given in three-quarter or full profile. By the later fourteenth century, these faces had entirely replaced the pagan god and goddess Sol and Luna in Plinian and other illustrations of the solar system.

The ubiquity of this facial imagery can be demonstrated by two crucial examples. In art, there is Giotto's famous fresco *The Last Judgment,* painted in the Arena Chapel in Padua. This painting depicts God the Father sitting in judgment and surrounded by angels, above whose heads hover the Sun on the left, painted in gold, and the Moon on the right, ashen in color with a distinct face on its crescent portion. Giotto did not entirely follow convention: his Sun is faceless and he did not confine the lunar face to the darker, concave portion of the crescent. Instead his image shows a gray, charcoal-like visage whose features vaguely suggest that they were based on the distribution of "spots" on the lunar surface. No verification of this can be made, however, given the lack of documentary evidence.

Among astronomical treatises, the second example was one of the most widely used textbooks between the thirteenth and sixteenth centuries. The two drawings of figure 5.1 are from a 1488 printed edition of John of Sacrobosco's *De sphaera* (On the sphere) and are meant to depict lunar and solar eclipses. Originally written around A.D. 1250, this brief work was a simplified introduction to spherical astronomy (planetary motions) and became a

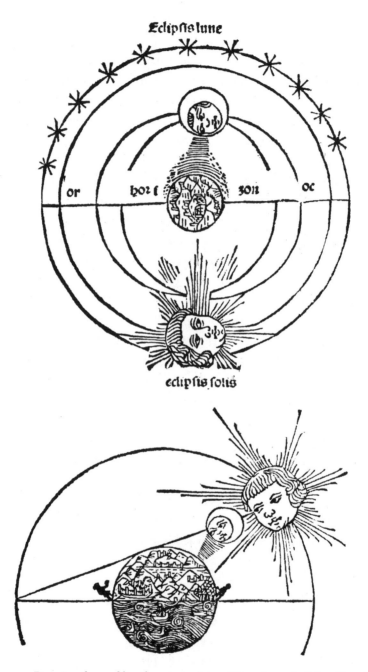

Figure 5.1. *Drawings from a fifteenth-century edition of* De sphaera *by John of Sacrobosco (thirteenth century), meant to illustrate lunar and solar eclipses.*

standard source in university courses down to the time when Kepler was a student in Tübingen. Earlier manuscript editions prove that the drawings of figure 5.1 are modeled on older versions, except that they give hair to the solar and lunar heads in naturalistic fashion. This not only departs from the strict allegorical tradition, but it also casts a degree of comical irony on it by literalizing the man-in-the-Moon image as a (decapitated?) form. The point, however, is clear: Giotto and Sacrobosco are neighbors beneath the light of the same lunar image.

For reasons as yet unknown, the face given the Moon was mostly male. This is difficult to explain, given the myriad connections between woman and the Moon, so much a part of Greek, Roman, and medieval culture and at least hinted at by the personification of the lunar orb in the form of a female goddess. Why were such associations erased at the level of representation? Evidence suggests that misogyny played no role in this change. Rather, a merger of ancient Greek and biblical folklore may have been involved, a complex transfer of the ancient notion of the Moon as a place of purgatorial atonement to biblical tales of punishment for sins committed. European folklore contains a number of versions of a tale based on the biblical story related in *Numbers* 15:32–36, in which a man who refused to rest on the Sabbath and instead gathered wood to sell was seized, bound, and stoned to death. Popular tellings have the man banished to the Moon, along with his sticks, which he continues to carry on his back. In some renditions, the offender is given a choice of whether to serve his exile on the Sun or on the Moon and he chooses the latter. In still different versions, the man is caught stealing cabbages on Christmas Eve; he and his booty are conjured to the Moon, where he can still be seen. Other European legends with medieval origins pose the man in the Moon as Judas or Cain burdened with a bundle of thorns, punished for his crimes by expulsion from Earth. The Cain image even gets a brief mention in Dante's *Divine Comedy* as a recognized folk belief (*Paradiso,* Canto II). Perhaps such stories also account for the tired and troubled expression so often given to the man in the Moon in medieval and Renaissance imagery. By the fifteenth century, the face had become an artistic cliché. So standard an image was it that even after Galileo's telescopic portraits of the Moon appeared in 1610, and painters in Italy and elsewhere began to depict the lunar body as a crystalline or cratered sphere, the man in the Moon continued to appear in serious works on astronomical and meteorological subjects.

A NEW SUNDERING OF THE HEAVENS

The massive influx of Greek-Muslim thought into Europe through translations mainly from the Arabic formed what has become known as the Twelfth-Century Renaissance,[3] one of the most crucial episodes in medieval intellectual history. At the same historical moment that the building of the great Gothic cathedrals took place, the foundations of modern science and the beginnings of the university system in Europe were solidified. It was at this time that the works of Ptolemy, Aristotle, Archimedes, Euclid, and others were introduced to Latin culture. These authors, however, did not come into Europe as they had left Greek culture more than a thousand years before. They had been copied, edited, corrected, partially rewritten, and illustrated by generations of scholars in Islam, who had absorbed their contents, debated their implications, and nativized their learning to a very different, evolving culture. What the West inherited was a much more complex alloy of Greek and Arabic learning. Along with Ptolemy came al-Farghani, al-Battani, Thabit ibn Qurrah, and al-Sufi. Aristotle and his version of the heavens went hand in hand with the commentaries of al-Kindi, al-Farabi, Ibn Sina (Avicenna), and Ibn Rushd (Averroes). Euclid was accompanied by the advanced trigonometry of al-Kwarizmi (whose name has succeeded to the present in the form of "algorithm"). And Archimedes arrived with interpretive companions too numerous to mention.

The effects on European astronomy were momentous. Change did not occur suddenly or dramatically; the basic position of medieval astronomy in society, its service to problems of enduring interest in a feudal, religious setting (calendar reform, time reckoning, seasonal predictions, and determination of holy days), did not vanish. What Greek-Muslim astronomy did, first of all, was to vastly improve the methods to achieve these tasks. It did something else as well, something of considerable relevance to our discussion of the Moon. The simultaneous introduction of Ptolemy and Aristotle effectively sundered the heavens in a new manner. Here we should recall the scheme of Geminus in the first century B.C., who wrote that it was the work of physics "to consider the substance of the heavens," and that of astronomy to study the "arrangement of the heavenly bodies . . . as well as their movements." Importing the basic texts of Greek-Arabic science and philosophy meant importing this ancient division, which had remained intact. As a result, the main current of astronomical study and observation turned toward the mathematical heavens, whereas philosophy and physics took over cosmological reflection, as well as discussions of planetary substance. Ptolemy's *Almagest,* for example, which was too difficult for the majority of students and scholars, was digested into two widely used introductory texts, John of Sacrobosco's *De sphaera* and the anonymous *Theorica planetarum,* both of which came with illustrations (fig. 5.1). These works appeared in the thirteenth century and effectively replaced Pliny and Martianus as the central textbooks of astronomy.

RISE OF THE LUNAR SURFACE:
A SAMPLING OF IDEAS

If *De sphaera* and *Theorica planetarum* were uninterested in the planets as actual physical bodies, such was not the case for Aristotelian cosmology and its commentators, of which there were soon an abundant number. These commentators were the ones who brought the "spottedness" of the Moon into discussions about the heavens. Plutarch's work on the lunar body does not appear to have been widely available before the Renaissance. Instead Latin writers drew upon Islamic authors, especially the twelfth-century philosopher Averroes, who in turn received much inspiration from the tenth- and eleventh-century scholar of Arabic science, al-Haytham (about whom we will have much more to say below).

One Latin writer who tackled the question of the Moon's "spottedness" was Alexander of Neckam (A.D. 1157–1217), reputed foster brother to Richard the Lionhearted and scholar of Greek and Arabic natural history. In his work *De naturis rerum* (On the nature of things), named after the tradition of works by Lucretius, Isidore of Seville, and Bede, Neckam informs us that some thinkers believe the Moon to be covered with mountains and depressions, whereas others interpret the dark areas as portions characterized by "greater obscurity." His own view was that the lunar spots were intended by God as a sign of original sin stained upon the heavenly bodies: only "when all the planets and stars shall stand as it were justified, our state too will become stable, and both the material Moon and holy church will be spotless before the Lamb."[4]

A very different type of image, bleached of such symbolism, was proffered slightly later by the consummate scholar of the late European Middle Ages, Albert the Great (A.D. 1193?–1280). Albert believed that the Moon was smooth and spherical, its dark areas caused by variation in the density or lightness of the lunar substance. This was the standard position of the Peripatetic school of Greek natural philosophy founded on the ideas of Aristotle. Albert, however, did not stop there, but offered a descriptive image for the lunar maria. Images of this type, as we noted earlier (see chapter 4), were known from antiquity: the apparent figure in the Moon had been seen as a cat, stag, lamp, drum, etc. In his *Meteorologica,* Albert produces a much more complex figure. Those, he says, who have closely observed the lunar disk in the early hours before dawn can discern a lion with its head on the east and on its back something resembling a tree trunk that thins to the west. Leaning against this tree, a man stands with his feet pointing toward the back end of the lion. Impressionistic as it is, the image indicates a high level of observation—indeed, it is complex enough to suggest that Albert may have sketched this arrangement of figures. Because Albert says nothing about the meaning of his image, it was probably a likeness, used for purposes of visualization and memory, stripped of mythic and religious connections. This type of secularism was new.

As noted, both Alexander of Neckam and Albert the Great were influenced by the writings of Arab natural philosophers. The most important of these with regard to the Moon and planetary astronomy was Ibn Rushd (A.D. 1126–1198), known to the Latins as Averroes, the most notable of the many commentators on Aristotle. Averroes, it appears, took some pains to derive a consistent theory of the Moon's substance and true nature from the various, partly contradictory statements found in Aristotle's work.[5] This was no easy task. Aristotle, as stated earlier (see chapter 2), had variable things to say about the Moon, noting that it "participates" in the Sun's light and is "as it were, a second and smaller Sun," yet it is also the origin of certain "creatures of fire" (such as the salamander) of terrestrial affinity.[6]

What does Averroes make out of this seemingly dual attribution? The Moon, he says, is similar to the Sun in an analogical sense, not a literal one. The relationship with the Earth, however, is more complicated. "The Moon has a relation with the terrestrial nature, because it is not luminescent"; those portions of the Moon "that are translucent, that do not glow by themselves . . . [have] a relation with the nature of water and air,"[7] which represent the nonluminous parts of the Earth. But what of Aristotle's stated link between the lunar substance and the element of fire? Different portions of the lunar surface have different natures, Averroes states; those more "translucent" and "obscure" bear a relation to the Earth, whereas those that are luminescent possess a nature in concert with that of fire and the stars. The Moon is thus, as in so many other formulations, a mixture of higher and lower matter, incorruptible celestial substance and imperfect "sublunar" material. The true essence of the lunar body, however, lies in light and visibility. Averroes makes this plain in an interesting way, drawing on a long tradition of Islamic deduction reaching back at least to the tenth-century polymath, al-Haytham. "It has been demonstrated," wrote Averroes, "that if the Moon acquires the power of lighting up from the Sun, it is not by reflection. . . . The Sun renders it luminescent first and then the light emanates from it in the same way that it emanates from the other stars."[8] A fluorescent lunar body allows Averroes to maintain further that the Moon is not rough and mountainous like the Earth, but smooth and perfectly spherical, as Peripatetic theory had stated. It is not a mirror, nor is it another Earth. It is a celestial ball of variable composition, whose dark areas define "portions of the surface . . . that do not receive the light of the Sun in the same way that other portions do."[9]

This is as far as Averroes goes. It is farther than Aristotle but way short of Plutarch. The Moon remains for "The Commentator" (as Averroes was known) a philosophical object more than a true physical one. Its surface is little more than the abode of certain essences, and in describing the nature of these, Averroes dispenses with any further interest in the observable features of the Moon.

AL-HAYTHAM: AN OVERLOOKED MASTERPIECE
ON THE SUBSTANCE OF THE MOON

Averroes claimed that his ideas on the Moon's ability to illuminate itself came from a near contemporary, the twelfth-century Hebrew scholar Abraham ben Ezra. It is clear, however, that both Averroes and ben Ezra drew their concepts on this topic, at least partly, from one of the greatest Islamic scholars of the sciences, al-Hasan ibn al-Haytham (A.D. 965–1039). Many volumes have been written about al-Haytham's contributions to science, especially in the areas of optics, mathematics, and astronomy.[10] It is said that, in addition to many other achievements, he was able to merge the deductive, often circular logic of Aristotelian natural philosophy with a far more inductive, experimental, and mathematical study of phenomena.

Al-Haytham wrote a great deal on the nature and behavior of light, and he applied his research to several treatises on the heavens. He wrote two works on the Moon, *On the Light of the Moon* and *On the Nature of the Marks Seen on the Surface of the Moon*. The latter work is the only true successor to Plutarch's famous dialogue prior to the Renaissance, a length of some fourteen hundred years. The book's German translator, Carl Schoy, characterizes al-Haytham's study as "the first significant step forward [since antiquity] toward a proper understanding of this lunar sphinx," and he is surely right. But Schoy, however inadvertently, is also partly correct in noting that al-Haytham's book has its limitations because its conclusions were "arrived at purely through optical considerations—since, at base, [to him] it was all a problem of optics through and through."[11] A problem of logic, too, we might add. When we read through this brief treatise, we find more than half of it devoted to debunking existing theories about the lunar "markings" and the remainder focused upon discerning the nature of these markings on the basis of extended deduction from a thin stratum of observations. At its outset, the work promises something unprecedented:

> If one were to carefully observe and consider the surface markings [of the Moon], one finds them to be of constant disposition, revealing no changes in themselves, neither in their form, their position and size, nor in their respective types of darkness. Superstitious men, and those who should not be taken seriously, have proposed their own, divergent opinions on this matter. Certain people hold that the spots belong to the lunar body itself; others believe that they exist apart from it, namely between the lunar body and the eye of the observer; still others conceive that they offer an inverted image [of the Earth], since the lunar surface is smooth and reflecting. . . . There are also those who maintain that the form of the earthly oceans can be seen there, in mirror image, while others say

that it is mountains and mountain ranges of our Earth that are reflected. Finally, there are some who believe that what is seen defines a [unique] form outlined by reflected rays falling upon the Earth. (pp. 1–2)

Al-Haytham's treatise is, therefore, like Plutarch's, a catalogue of existing ideas about the lunar surface. Also like Plutarch, this author intended to dispense with most of these ideas. Whereas the Greek author selected one particular concept for applause from among the many discussed, al-Haytham had his own to propose. Doubtless Arabic scholars were directly and indirectly influenced by Greek ideas on the Moon; some of the beliefs discussed by al-Haytham (e.g., the Moon as a reflecting sphere) were current in Hellenistic philosophy and can be found in Plutarch's own work. Does this mean that al-Haytham used Plutarch as a source? The possibility certainly exists. A similarity in titles alone might suggest as much. Beyond this, however, some of the same reasoning and evidence appear in both works. Yet the conclusions reached by these two authors about the nature of the lunar "spots" are entirely different and, in fact, mutually exclusive. In matters of style, too, Plutarch and al-Haytham share very little. Whereas the Greek author chose a dialogue format and thus could include a range of specific rhetorical voices as well as humor, literature, myth, and philosophy, al-Haytham (if we are to accept his translator's version) wrote in a consistently flat, no-nonsense style entirely reminiscent of Aristotle. His book is much briefer (a mere thirty-four pages in its published translation) and less entertaining. Here lies a crucial point: it is distinctly *not* a literary science that interests al-Haytham, but a philosophical and observational one. The Moon to this author is not a subject for imaginative flights of fancy or folly; it is a very real object whose essential nature must be discerned from the workings of clear-sighted logic.

What were some of the specific "theories" that al-Haytham sought to debunk? First mentioned and apparently among the most popular were those that posited some intervening substance in space itself, whether between the Moon and the Earth or the Moon and the Sun. Such a substance might include a certain "moistness" or a series of "vapors" drawn up from the surface of the Earth (again, a Greek idea, which is also found in Pliny). Alternately, denser or more "opaque" areas hovering between the Moon and the Sun were thought to partially block out the light received by the lunar surface, thus creating the appearance of "spots." These theories, therefore, were based on the idea that the Moon shines mainly by reflected light. A second group of hypotheses proposed that the Moon is covered by a thin, transparent layer and that either this layer is itself spotted or else acts as a complex stratum in which the light from the Sun and the light from "behind (within?) the Moon" mix and partially cancel each other. Al-Haytham also discusses and discounts the concept that the Moon's appearance is due to topographical roughness that throws shadows on what is otherwise a smooth and crystalline surface. Such roughness, he states, was thought by some to result from elevated terrains, such as mountains, and by others from low areas, such as cavernlike holes.

Considered closely, these theories do not merely represent different accounts of an identical observation. To see the lunar maria as "clouds" out in space, as pure reflections of earthly oceans, as embedded "darknesses" in a translucent film, or as shadows cast by mountains or valleys requires a difference in specific cognitive emphasis. This diversity in cognitive style tells us that there were no standards at the time, as in antiquity, for observing the Moon. No central guide existed for making sense of the seen. Such guides *were* present with regard to the motions of the planets and the Sun, the visual order of the constellations, and the geometric relationships of the solar system. These guides were mathematical and, in the case of the constellations, also allegorical. They did exist, however, and helped to standardize perception. This was not true of the Moon. Lunar perceptions remained diverse, unsettled, even fanciful up to the time of Galileo. As we shall see, it was not the telescope per se that finally, after so many centuries, set down the basic rules of the Moon's visual reality; it was Galileo's own magnificent use of this instrument to produce convincing images of witness that established a true perceptual rhetoric for the lunar surface.

Galileo, in a manner of speaking, recreated the Moon in the image of the Earth in an entirely literal fashion. Any of the visual concepts that al-Haytham discussed might also have been tied to terrestrial realities, such as clouds, glass, crystal, or translucent stone, all of which could have been used to support the perception of the lunar surface as embodied light. As for the vision of mountains and caves on the Moon, obviously it is the Earth itself that is used as a perceptual model, just as it was in Plutarch's day. Yet none of these concrete, everyday analogies are invoked in al-Haytham's text. Everything remains on a more elevated, philosophical plane. In writing of the lunar surface-as-landscape theory, for example, the author never uses terms such as "hills," "valleys," "mountainous," or "rolling"; he speaks only of "roughness" (Rauheit) or "unevenness" (Unebenheit). The only comparison made is an indirect one between the Moon and a certain type of optical "body."[12] This lack of earthly comparisons is just as true for Averroes, whose work appeared several centuries later. This point may seem minor, but it gains much meaning when compared with later discussions of the lunar surface during an era of increasing naturalism in art and science in Europe.

Without going into detail, al-Haytham dispatches most ideas about the Moon on the basis of two fundamental principles. The first of these stems from observation and was also used by Plutarch: the spots on the Moon cannot be pure reflections, or the result of space vapors, or anything similar because they would naturally change shape, size, and location at different times of the lunar cycle and from different positions on the Earth. The "spots" on the Moon never vary, therefore none of these theories can be correct. Al-Haytham's second principle, however, is one he deduced from optical considerations of the Moon's light and origin, as put forth in his earlier work *On the Light of the Moon*. This principle denies that the Moon shines by reflected light alone, maintaining that its luminosity is due to a combination of reflection and the native capacity of the lunar material to absorb and emit, or "give back," a certain quantity of light received from the Sun. "The Moon," writes al-Haytham, "becomes self-illuminating when struck by sunlight" (p. 11). "That light produced by each point of emanation is called 'secondary light,' while the light reflected from the smooth body [of the Moon] is called, analogically, 'primary light.' Secondary and primary light are produced together, when the smooth body is reflecting in the line of sight" (p. 13). It is this idea of a self-illuminating Moon that Averroes adopts in his own treatise, although taking it to a more extreme position.

How, finally, does al-Haytham explain the markings on the Moon? First, he concludes on the basis of the theory just noted that differences in the light given off by lighter and darker areas of the Moon indicate that "the body of the Moon [must] be entirely different in those places where spots occur" (p. 20). More specifically, "the entire lunar body possesses the power to absorb light; only in the places where spots occur is this power incomplete, and this because of some impediment" (p. 30). The origin of this impediment, we are eventually told, is that the Moon, like other heterogeneous bodies, contains certain areas characterized by an increased "density" *(Dichtigkeit)*.

At this point, the treatise ends. In modern terms, we do not seem to be left with much; but in fact, the author had gone beyond most of his contemporaries. The secret of the Moon's spots, he implied, the answer to this ancient "lunar sphynx," was not to be found in phenomena of light after all but in the nature of the Moon's physical substance. Hints of the idea of "dense" and "rare" lunar substance existed in antiquity and were certainly later found in the writings of Arab scholars such as Ibn Sina (Avicenna). Al-Haytham, however, gave this concept one of its first major articulations. In this regard, something else this author says is also noteworthy: "If what we perceive in the lunar surface is only reflected light, then it can be maintained that the spots on the Moon represent nothing other than physical irregularities in this surface, which hinder the reflection of light" (p. 11). Inverted, this argument states that the Moon must have a significant topography if it can be shown to shine by reflected light.

What we find lurking in al-Haytham's little book is very similar to what we saw in Plutarch's much larger one: an essential interest that strikes near the heart of the lunar appearance, extending itself into questions and statements about the actual physical reality of the Moon. Granted, Plutarch is more explicit; his "mountains flaming bright" affect our imaginations deeply and immediately. Both authors, however, succeed in drawing the reader's attention to the lunar surface itself. This is where al-Haytham's narrative finally lands us, despite its philosophical-optical baggage. His text even resembles the very body he wishes to investigate: thick and opaque in some parts, brilliant and unobstructed in others. If Averroes refused to acknowledge any debt to him, it was perhaps to avoid any direct comparisons.

THE MOON OF THE FOURTEENTH CENTURY

By the early 1300s, the ideas of Aristotle and Averroes were in wide circulation throughout European intellectual culture and commanded enormous, if sometimes conflicting, allegiances. Although it does not appear that fourteenth-century Scholastic philosophers knew of al-Haytham's work on the Moon (there seem to be no Latin versions of it), they nonetheless repeated much that was in it through their debates and discussions on Averroes. Contrary to what has often been said, however, European authors did not merely repeat The Commentator's words. One example of this, as well as a sign of how generally known were Averroan concepts, is the discussion of the lunar substance that appears in two works by Dante, his *Divine Comedy* and the unfinished *Convivio*. In both works, Dante mentions that it is a widely held belief that the dark portions of the Moon represent "rarer" material and the light areas "denser" material, which is a complete inversion of Averroes's own position.

Not content with this explanation, Dante has Beatrice refute it in the *Divine Comedy* by proposing an actual experiment along Aristotelian lines (involving mirrors and a candle to show that a reflected image does not become darker with distance) and then replaces the dense/rare idea with a diffuse, Neoplatonic scheme dependent upon angels governing each of the planetary bodies and a single "mover" who is variably expressed in them.[13] What is significant about Dante's scheme, however, is its proposal that the Moon is made up of more than one substance: "[T]he which in quality, as well as quantity, may be observed of diverse countenance." This was a much more radical departure from reigning interpretations of Averroes, which preferred a more or less homogeneous lunar body. It was also not well accepted within the precincts of Scholastic philosophy, even among such influential and innovative thinkers as John Buridan, Nicholas Oresme, and Albert of Saxony.

John Buridan (ca. A.D. 1300–1358) wrote a treatise on Aristotle's *Metaphysics,* in which he also commented directly upon Ptolemaic astronomy, following Averroes's denial of the existence of epicycles. Buridan speaks only briefly of the Moon and its appearance, when discussing whether an epicycle should be assumed for its motion. It should not, he writes, "because then it would follow that in the spot of the Moon which appears as if it were an image of a man whose feet always appear to be below [or toward the bottom], the feet would sometimes appear above [in the upper part of the Moon]."[14] This image is less complex and was probably much more prevalent among scholars than that of Albert the Great. Yet it shares with his notion a more purely descriptive component. Nothing of this attempt to identify a figure in the pattern of light and dark on the lunar surface exists in the work of Averroes. Buridan and Albert both treat the Moon as a single body with a single substance but also speak of it as a picture to be painted in words.

Between A.D. 1370 and 1377 Nicholas Oresme translated Aristotle's *De caelo* into French, giving it the title *Le Livre du ciel et du monde* (Treatise on the heavens and the universe). In this work he included a commentary in which he made many statements about the Moon. Oresme mentions Albert the Great's lion/tree/man image, yet he also cites Averroes in many instances. He states definitively that "[t]he moon is a perfectly polished . . . transparent and clear body such as crystal or glass."[15] Its "spot" or "shadowy figure" results because different parts of the lunar substance are "transparent and clear" to different degrees, the optical equivalent of "dense" and "rare." The author also spends some effort refuting competing ideas about the Moon's surface, especially the classical notions that it is a mirror of the Earth or obscured by "heavy vapors" lying between the Earth and Moon that are attracted by the "cold body" of the latter. (This was a popular idea among Islamic scholars of Greek science.) At some point, however, Oresme makes a curious comparison: "It should be noted that, in the case of an alabaster stone, those veins or sections that are most clear and through which one can see almost as clearly as through crystal seem darker and less white than the other parts; the same is true of parts of the Moon. Thus the clearer some parts are, so that the Sun's penetration is deeper, the darker those parts appear" (p. 459).

A piece of terrestrial rock, an ornamental one at that, becomes a model to explain the nature of the lunar surface. In these few lines, Oresme introduces a new world of immediate, visual evidence that one can physically hold before the eye. For the first time since antiquity perhaps, literal pieces of the Earth are proposed as the key to the nature of the Moon. Oresme's work has more than a few concrete analogies: the fixed stars are said to move "like a nail lodged in a ship" (p. 453), and the heavens cannot be divided "as one divides a wooden log" (p. 455).

One last fourteenth-century author who deserves mention here is Albert of Saxony (ca. A.D. 1316–1390), whose work, *Questions Regarding the Heavens and Universe,* addressed the problem of lunar substance in a manner that, at first, seems an almost word-for-word citation of Averroes:

The Commentator issues [an] opinion, which I believe to be true. The spot [on the Moon] issues from the diversity of the parts of the Moon. . . . The parts in which the spot is seen are the rarest, which renders them least capable of glowing. The parts next to them are the densest, and because of it, they glow most. . . . The Moon is simple in substance, in fact, but that would not prevent it from exhibiting differences in density and rarity between its various parts.[16]

Like Dante's protagonist in the *Divine Comedy,* Albert of Saxony reverses Averroes on the lunar maria and their cause while upholding the idea that the Moon's substance is "simple," i.e., homogeneous. Albert is less in touch with the growing preference for the concrete than are his relative contemporaries, John Buridan and Nicholas Oresme. His Moon is very much a Scholastic orb, unmediated by the new comparisons and analogies that seek to explain it in entirely familiar terms and that would continue to gain force in the century that followed.

THE RISE OF NATURALISM: ART AND SCIENCE UNITED

Late medieval art began a profound shift in its aesthetic priorities with respect to the natural world. In the place of stylized, wooden, or gestural portrayals, expressive of an art that served as didactic surface, a new type of realism intruded. By the mid-thirteenth century the capitals of cathedral columns in France; the margins of French, German, and English book manuscripts; and the illustrations of most herbaria and other scientific works displayed the closely observed features of plant species, insects and birds, known and mythical animals, and Christ with wounds that gushed or flowed.

The ascent of naturalism has been a topic of frequent scholarship ever since the early decades of the twentieth century.[17] This ascent appears to have begun in sculpture and progressed from there into painting, book illumination, and other expressions. So striking is this change from previous Romanesque styles, so eager is it for the world of appearances, that it has often inspired flights of exaggerated discovery: "The thirteenth century sculptors sang their *chant de moi,*" writes one well-regarded authority. "All the spring delights of the Middle Ages live again in their work. . . . The [late medieval period], so often said to have little love for nature, in point of fact gazed at every blade of grass with reverence. . . . It was these breeders in stone, these Burbanks of the pencil, these Darwins with the chisel, who knew nature and had studied botany and zoology in a way superior to the scholar who simply pored over the works of Aristotle and Pliny."[18] The truth, however, requires that this observation be amended. Not only were these early naturalists steeped in the writings of Pliny, but they were also probably acquainted with the herbals and bestiaries of Arabic writers, the writings on animals by Aristotle, and the physiology of Galen, all so recently translated and so clearly superior to similar works by earlier Latin authors.

The new feeling for natural detail came from many things, not all of them well understood. Beginning mainly in France, it was tied to the burgeoning court of Paris, with its growing love of learning and finery. The medieval world had become a vastly larger place than it had been in the time of Charlemagne, spurred forward by the growth of towns and international commerce, the rise of the artisan classes, the spread of literacy, the advent of pilgrimages and crusades, and technological advancements.[19] A potent new mobility had been added to European society by the late twelfth century— mobility of objects, goods, currency, persons, books, languages, words, and images. Partly as a result of these expanding circulations, an eager materialism became inevitable. It was aided by religious reforms such as those of the Franciscans, who saw and praised God in the "littlest of things." No longer was the physical world a storehouse of symbolic images, "an aesthetic expression of ontological participation" in the play fields of the divine.[20] While retaining such universality of the *tableau sacré,* it had also become an atlas of objects, a "book written by the finger of God," as noted by Hugh of St. Victor (A.D. 1096–1141).[21]

No more remarkable example of the new piety of naturalism exists than in a text by this same author comparing holy meditation with green wood catching fire. Hugh's description of the flame's struggle with the log, the clouds of swirling smoke, the quality of flickering light, and the progress of combustion itself go on for more than a page. They are delivered in a vivid, dramatized prose; the author is not merely describing things for us, he is telling a story, painting a picture. His observations bear us along, beyond the edges of any original comparative dimension, into a scene of wholly material process.

If the fire first seizes hold of the green wood with difficulty, it soon does so with ever stronger gasps, flaring up against the exposed pith. One sees how the thick, dark clouds of smoke rise and envelop the yet measured glow that barely shines within, until, gradually, the flames more fully awake, smoke and darkness disappear, and the pure glow of fire emerges into the foreground. Now the victorious flames take command and spread themselves over the entire burning mass, like a funeral pile . . . flickering here and there, penetrating and shooting forth from the victim material. Only then, when it has penetrated the innermost portions and drawn all into its power, does the fire grow quiet . . . and every sound softens. The raging and devouring flame becomes still and peaceful, having forced all to submit and be incorporated into its sympathy, finding nothing any longer that might be alien or opposed to itself.[22]

Reading such words, we sense a mind caught up in its own ability to observe, record, and dramatize, a mind in love with the narrative production of images.

Between the middle thirteenth and late fourteenth centuries, the regional base of the new naturalism shifted from holy to secular workshops. Commissions offered by and through the Church were no longer the staple of artists and illustrators everywhere; many now found patrons in the court of the French kings and later in the metropolitan centers that bloomed under the wealth, power, and taste of the brothers of Charles V. These dukes of Berry, Burgundy, Anjou, and Orléans acted as the benefactors of the burgeoning arts and sciences and did so in sumptuous fashion. The greatest artists of this period, up through the first several decades after 1400, were Flemish and Dutch by birth and often by training. The most famous are Claus Suter from Haarlem; the Limbourg brothers from Guelders; Melchior Broederlam from Ypres; the Master of Flémalle and Jan van Eyck, both from the Netherlands. All of these artists took the "return to nature" greatly beyond anything previously conceived. They produced a scholarly art, steeped in textual sources (not only the Bible but also the works of Pliny, Discorides, Aristotle, and Galen), that was also worldly in the most immediate sense. Quite suddenly the artistic capacities of antiquity, for so long the standard model of *renovatio,* were left behind. For the first time, it appears, artworks became the stage for revelations of visuality and displays of collected observation. Stone, tempera, and oil were transformed from tools to media, from implements for expressing spiritual universals to visual textbooks in which both common and uncommon objects of this world could be documented with photographic precision.

This is not to say that these artists left behind all traces of medievalism. On the contrary, their training and the books upon which their scholarly sense of the world were based came directly out of medieval sources. Some more than others retained crucial aspects of past aesthetic tradition, and all kept something of the earlier centuries intact, whether in the gestures and arrangements of their holy subjects, in the use of gold to depict light, or in the lack of secular topics. But here their debt ends. A difference far greater than mere application lies between oak leaves and animals crawling across a sacred page and vast landscapes populated by trees, rivers, valleys, mountains, famous castles, ships, and every sort of human dress and face. The new materialism in art carried with it a necessary precision, an adoration for the physicality of things. It was a monument to observation. It is no accident that the first naturalistic drawings of the heavens occurred outside the limits of astronomy itself. It is a vast step in the history of illustrating the world that the Moon and the stars finally entered the list of worthy subjects at this time.

6

The First Drawings of the Lunar Surface

ART ASCENDS THE HEAVENS

There is a small yet striking truth about medieval European art: for more than eight hundred years, from the end of antiquity to the opening of the Renaissance, not a single figure was painted to cast a distinct shadow. After so long an absence, shadows reappear in the frescoes of Masaccio and in the oil paintings of early Netherlandish artists, most notably Jan van Eyck.[1]

For Masaccio, whose *Acts of the Apostles* appears in the Brancacci Chapel of St. Maria del Carmine in Florence, shadows are a central character to a particular drama—the tale of St. Peter healing the sick with the touch of his shade. The artist has attempted to reenact events for the viewer, who is then a virtual eyewitness. There is a crucial continuity with the medieval universe here. From St. Gregory onward, one of the stated functions of art was to teach the illiterate the stories of the Bible.

For Van Eyck, on the other hand, the "conquest of appearances" has a very different quality to it. Shadows, for example, are never magical or miraculous. They are instead ordinary, simple, darkening extensions of objects in the everyday world. A small curl in a gown, a loose shoe on the floor, the gentle press of a finger into flesh, a candle lit against a wall: such commonplace occurrences are among the phenomena that call forth the artist's consummate effort at shadow making. If Masaccio was interested in gesture, Van Eyck was interested in specimens.

THE LIMBOURG BROTHERS: A VERY RICH SKY

The ceilings of some cupolas in Roman architecture, painted with the images of celestial gods and goddesses, are also decorated with stars, usually in white or gold.[2] This classical motif continues in medieval Christian painting of the heavens down to the time of Giotto, who used it to cover the vaulted ceiling of the Arena Chapel in Padua. Meant to serve ornamental purposes, the stars in the Arena Chapel are a flat gold (the medieval equivalent for light), with spikes radiating outward from a central dot. They are spaced regularly in a deep blue medium, suggesting a literal rendering of the "vault of heaven."

Such stylization is decidedly not the case for a striking image painted by one of the Limbourg brothers (presumably Paul) as part of the famous series of miniatures, *Les Trés Riches Heures du Duc de Berry.* The brilliantly colored and meticulously executed images of this series, which together provide nothing less than a wholesale archive of daily life in late medieval times, were worked on by the brothers themselves mainly between 1405 and 1416, and thereafter by their followers into the 1440s. The Limbourg images are sumptuous, learned, and mundane and treat many more subjects than those shown in the well-known twelve calendar scenes that grace the pages of most art history texts.[3] In a number of paintings, the Limbourg brothers followed tradition and decorated the heavens with six-pointed stars randomly scattered in a matrix of cobalt blue. This is true of the calendar scenes (which also depict the Sun as a great fireball in the hands of Sol, who drives his wagon through the clouds) and of a few religious images (e.g., St. John on Patmos). Another miniature, however, goes far beyond such conventionalism.

This image, entitled Hours of the Passion, depicts an event in the Gospel of St. John (18:3–6) during which a group of soldiers led by Judas have gone in search of Jesus during the night. Approached by Christ himself, with Peter at his side, the soldiers mysteriously fall back and collapse on the ground in a helpless heap. The painting shows Christ and Peter standing over a pile of helmets, bodies, and lances, framed by a brilliant evening sky. Everything is shown in dark, shadowy tints, providing a remarkably realistic sense of night. The only light issues from a halo around Christ's head, a fallen lantern and two torches, and the azure sky filled with a myriad stars and three meteors, their tails in streaks of gold. The stars are of many different sizes, ranging from pointed varieties to tiny blebs, and are realistic enough in their distribution, brightness, and placement to suggest a panoramic view of the southeastern heavens in February, bedecked with the constellations Aquila, Lyra, and Hercules. Closer inspection reveals that such identification is questionable; the major star groups are not precisely where they should be, and "extras" appear to be thrown in for good measure. Yet the attempt to render the impression of an actual nighttime view is unmistakable. This scene is one of the first attempts to create a truly naturalistic effect for the heavens at night, complete with astronomical phenomena in the form of heavenly signs—shooting stars—a *trés riche* sky by any standard of the time.[4]

What of the Moon, then? In another night miniature, the lunar body appears. As might be predicted, it is an image of the Crucifixion, denoted *The Death of Christ*. In the upper left corner of the painting hangs the Sun, a faint gold, and in the upper right, a faceless crescent Moon, painted not gold but white and showing the outline of the entire orb. In the upper right margin is a cameo depiction of an astronomer with books of tables (ephemerides presumably) on a stand before him, his eyes raised to the sky in an attempt to comprehend the incomprehensible. The addition of this small portrait changes the overall meaning of the image significantly: instead of a pre-ordained and epochal event, Christ's death is now an occurrence that generates miraculous happenings in heaven, confusing to even the most advanced science.

These paintings of the Limbourg brothers are dramatic in the narrative sense. Like Masaccio's fresco of St. Peter healing the sick, they tell stories. Their use of nature as a stage for this purpose was apparently intended to animate these tales so they might hold the eye more deeply. This is true of all of the work in *Les Trés Riches Heures*. This purposeful use of nature in art became a tradition in France during the fourteenth century, first under the patronage of Philipe the Tall, then under Philipe VI, both of whom had employed the services of Jean Pucelle (d. 1332), the most renowned painter of his day. Pucelle was among the first group of artists to profit by the transfer of patronage from Church to court. Under the sponsorship of the new Valois dynasty, beginning with Philipe VI (who ruled 1328–1350) and his son, Jean le Bon (1350–1364), a heavy favoritism came to rest upon the arts, which drew many painters and poets to the French royal scene in Paris. The outlook of the court was grandiose, self-elevating, international, and highly learned. Artists were encouraged to produce images that reflected the beauties and details of France—France as a setting for modern history, for the birds and beasts of the hunt, for the flora and fauna of Europe, for castles and cathedrals, and most of all, for holy tales of the Bible.[5] Jean, Duke of Berry (d. 1416), second son of Jean le Bon, inherited this general outlook from his father. A lover of manuscripts, art, and music, he commissioned many Books of Hours, of which the version by the Limbourg brothers was the last and most impressive.

That the stars and the Moon should appear in the new art of the fourteenth century is no surprise. That this stage of artistic conception should be utterly surpassed, however, and brought to a height unimaginable in either France or Italy is less to be expected. The final step in the visualization of the heavens prior to the invention of the telescope came at this same time, during the first decades of the fifteenth century. It is a step that has not yet been appreciated in full, either by historians of art or those of science. Yet it represents the moment when a new universe enters the realm of expression.

THE FIRST LUNAR PORTRAITS:
JAN VAN EYCK AND THE ART OF OBSERVATION

The lunar face, as well as a large portion of the Earth's surface, finally became part of Western representation in the work of Jan van Eyck (1385?–1441). Van Eyck, the early Netherlandish master responsible for perfecting the use of oil paints to convey the subtlest effects of natural appearance, worked in the employ of Philip the Bold, Duke of Burgundy and nephew to Jean du Berry. Van Eyck's command over realistic technique, so striking today, was no less legendary in his own time. What he achieved for the history of art remains a source of endless controversy.[6] But his importance to any history of observation—in particular, the recording of the physical world as a documentary act—can hardly be debated.

It is precisely this aspect of Van Eyck's work, its precision in rendering the external, nonhuman, nonorganic universe, that has gone most unnoticed. Scholars in the history of astronomy, for example, have until very recently attributed the first naked-eye drawings of the lunar surface to Leonardo da Vinci, specifically to three sketches that appear in his notebooks.[7] These drawings include two rough pen-and-ink outlines showing a jack-o'-lantern-type man-in-the-Moon image and a third, more careful charcoal sketch of the western lunar half. The first two sketches were produced about 1504–1505, whereas the third was drawn sometime later, about 1513 or 1514.

A painting completed by Jan van Eyck nearly a century earlier, however, confirms that any priority for the first true image of the lunar surface must be lifted from Leonardo's hands (fig. 6.1).[8] This work, fittingly enough, is *The Crucifixion* (1420–1425), the left half of a diptych that also includes as its right panel *The Last Judgment*. The artist, perhaps in collaboration with his brother, Hubert, also produced four other versions of the Moon. One of these is a waning gibbous in the *Knights of Christ* panel of the famed *Ghent Altarpiece* (1426–1432); another, showing the late crescent phase with earthshine, appears in the unfinished *St. Barbara* (1437; probably by Jan alone); a third, very faint crescent image appears in *Rolin Madonna,* just above and to the right of Chancellor Rolin's head; and a fourth, magnificent three-quarter Moon image, with all of the maria faintly visible, is part of a book-of-hours miniature for the Hours of Turin, entitled *Birth of St. John the Baptist* (variably dated before 1417 or after 1430). Of these works, the last is the most hotly contested regarding Jan van Eyck's authorship; it has been alternately attributed to his brother Hubert, to an unknown early master, and to a student of Jan.[9] The Moon would appear to clinch the case—or at least to add fresh tinder to the debate—because no other artist of this time period, whether in the Netherlands, France, Spain, or Italy, ever portrayed the lunar surface, let alone with such precision. All five works show the Moon in broad daylight. In the *Knights of Christ* panel, it appears below a precisely executed cloud formation and above the head of St. George, in a hazy sky with the maria mostly obscured. In the *St. Barbara* and *Rolin Madonna* images, the Moon is a thin crescent (a one- or two-day-old Moon), just above a V-shaped flock of

Figure 6.1. The Crucifixion *by Jan van Eyck (painted 1420–1425), showing the first naturalistic representation of the lunar surface in Western culture. Reprinted by permission of the Metropolitan Museum of Art, Fletcher Fund, 1933 (33.92a).*

geese in the former and nearly hidden in a light formation of cloud in the latter. The full orb is plainly outlined in both paintings, with the dark portion faintly lit—a naturalistic portrayal of earthshine. The tiny chalklike disk of *Birth of St. John the Baptist* gives us a better quality image of the lunar face, low in the sky, flecked with the maria that emerge more fully under slight magnification.

It is in *The Crucifixion,* however, where the Moon's surface hangs most strikingly before our eyes (fig. 6.1). This image, though obedient in general staging to traditional depictions of the event, overturns the iconography that had been in place for a thousand years. Instead of a miracle in the heavens drenched in symbolism, Van Eyck gives us a terrestrial sky flooded with sunlight, beautifully rendered clouds, and a chalk-colored Moon low on the horizon. This orb, moreover, instead of retaining its iconographic position above and to the right of Christ or else flanking his head, is placed below an arm of the cross to which one of the two thieves is bound (fig. 6.2). It is a gibbous Moon, and on it the major lunar maria are all shown as they would appear in the late afternoon, perhaps corresponding to the "ninth hour" (about 3 P.M.) in the biblical account. Van Eyck's Moon is small, the same size as the heads of the people (about 0.5 inches in diameter); it rests by itself, alone in one portion of the sky, as if on a slide for microscopic viewing, just above a group of snowcapped alpine mountains crossed by late afternoon shadows. The terminator (border of the shadowed portion of the Moon) is drawn with visible, naturalistic irregularity, being inclined about thirty-five degrees to the horizontal. A detailed view (fig. 6.2) proves that the maria can be identified in two major trends, one including Maria Serenitatis, Imbrium, Fecunditatis, Tranquillitatis, and Nectaris, and the other trend below including Oceanus Procellarum, Mare Nubium, and Mare Humorum. In particular, the Mare Fecunditatis is positioned very close to the western edge and the Mare Crisium is either partly in the night hemisphere or so close to the edge as to be unrepresentable. Such observations suggest that a libration (position of the Moon within its longitudinal shifts) might be estimated.

The naturalism of Van Eyck's image is overwhelming, all the more so in that it rejects with unqualified detail so many centuries of Crucifixion representation. As we noted previously, the Limbourg version shows the world cast into darkness, with both Sun and Moon present as background objects. The impossibility of the scene is central to its depiction of divine intervention; it

Figure 6.2. Detail from figure 6.1. Reprinted by permission of the Metropolitan Museum of Art, Fletcher Fund, 1933 (33.92a).

employs naturalistic technique toward a rendering of the biblical miracle. Nothing could be further from the qualities of Van Eyck's image, where the everyday physical universe takes center stage. We see clouds, hills, and the rock-strewn ground posed as a specific location at a specific moment. The mountains in the painting lead forward to a broad, meandering river, then a castled town (presumably Jerusalem). In the foreground below the three crucified figures is a scene inhabited by all manner of onlookers, each one dressed in an entirely different set of apparel and holding a separate pose: some are gazing upward; many are laughing, telling jokes, or engaged in conversation; others simply jeer or stand uninterested in the scene. If not for the weeping Madonna, comforted by a small group in the lower portion of the painting, and the written words above Christ's head, there would be scant indication that the scene represented anything other than a fairly ordinary execution. Indeed, the pockmarked, chemically weathered limestone beneath the Madonna's feet or the gleam of a rider's sword receive fully as much attention from the artist as the features and figure of Jesus.

The Moon itself, therefore, is part of a much larger canvas of objects, a veritable museum of collected observations lacking any obvious overlay of divine touch. The lunar face is a single entry among a multitude of careful, direct visual studies, which include many aspects of the inorganic world. It is with Van Eyck that late medieval naturalism, previously focused upon the organic universe of feudal experience, is broadened to include this physical cosmos whose details are presented with a fidelity that far exceeds anything achieved by the whole of the Italian Renaissance—a period for which such things as rocks, rivers, mountains, and so forth were most often elements of staging. One indicator of Van Eyck's unprecedented, and for centuries unrivaled, naturalism is the very high degree to which his world can be investigated by contemporary scientific perception and methods. This has been done, in fact, in at least two cases with regard to weather and rock formations.[10] In general, Van Eyck's strata and landforms permit geologic analysis; his clouds, meteorological forecasting; his topography, geographic investigation; his Moon, astronomical inquiry.

That this painter far superseded the intellectual burdens and limits of his own time raises the question of his motives. That his eye saw and his hand recorded realities overlooked by his contemporaries makes one eager to probe the origins of such capacity. These origins, however, must remain the subject of speculation. Scant biographical information and the lack of any personal writings have long ensured this must always be the case. Van Eyck left behind no notebooks like those of Leonardo, indeed no writings at all. Such lack of evidence is typical of early Netherlandish artists and seems to reflect their situation as court painters: they were documenters, not thinkers or theoreticians. Their paintings were their texts. This is perhaps more true for Van Eyck than for other artists because of his tendency to include many images of books in his paintings (*Ghent Altarpiece,* for example, has more than fifteen) and more directly, because of his unique proclivity for actually signing a number of his works in one fashion or another. His most common signature, offering a sort of personal motto—"*Als ich can*" ("As I can")—was sometimes written in Flemish, other times in Latin, and even on occasion in a mixture of Greek and Roman letters. The painter thus considered writing itself a source of visual fascination, part of the world of the seen. With its mingling of rhetorical modesty and obvious pride, his motto speaks not only of superior interest in the material universe, but also of devotion to the virtuosity of recording its appearances, the optical surfaces of experience.

As mentioned previously, Van Eyck poses his material universe as if it represented a specific location at a specific moment. This is part of the illusion; in fact, he did not create literal documentaries of particular settings on particular days. The Moon, as any astronomer might note, can not occupy the position in the sky shown by *The Crucifixion* (see note 10). Moreover, the temperate climate indicated by the distant mountains and green valley of the river is incommensurate with the scrubby barrenness covering the hill of Golgotha. Van Eyck, as we noted, was a collector, a creator of dioramas. Each of his paintings is a kind of *Kunstkammer* ("cabinet of wonders") of precisely rendered observations from various places. The importance to science of such an approach should be clear: liberating the object world from stylization imposed by religious demands and Scholastic final principles has often been described as one of the fundamental advances either of the Renaissance or of the Scientific Revolution. Van Eyck offers proof, however, that this freeing up of the natural world actually took place earlier, partly *as a result* of religious feeling—religious feeling, we should say, that was merged with or tempered by erudition and lush materialism.

Van Eyck's familiarity with the knowledge of his time is fully apparent in his love of books; the studied realism of his jewelry, buildings, plants, trees, musical instruments, and armor; his precise imitations of earlier medieval drawings (evident in the floor tiles of his *Annunciation*); the cosmopolitan variety of clothing his people wear; and much more. At the same time, Van Eyck's paintings have often been interpreted as gardens of medieval iconology, overgrown with religious symbols that presumably ripen into contemplative mystery, silence, and enigma.[11] "Nothing is as it appears to be," writes one recent scholar in this vein, expressing both the possibilities for frustration and the endless latitude for interpretation that such an approach allows.[12] Yet this does not seem convincing, given the display of virtuosity devoted by this artist to reproducing exactly what "appears to be." It has also been suggested that the painter was practicing his art as a special type of spiritual fidelity, based on a belief in the material world as a scroll of God's truth written in the Creation, with every surface worthy of attention for having attracted the care and effort of the Creator.[13] Van Eyck's achievement has also been placed in a larger intellectual context, specifically in relation to "that nominalistic philosophy which claimed that the quality of reality belongs exclusively to the particular things directly perceived through the senses."[14]

This seems a more palatable interpretation, yet one largely divorced from Van Eyck's own real-world position as a court artist, with all its attendant demands. Van Eyck was hired to work as advisor and chamberlain to Philip the Good, duke of Burgundy, a well-educated humanist noble, procurer of art objects and scientific literature, lover of conspicuous pageantry, and one who surrounded himself with men skilled in various arts and disciplines. Under the Duke's favor, Van Eyck appears to have been encouraged to pursue his many interests in goldsmithing, alchemy, architecture, and town planning, as well as the chemical technology of mixing paints and oils. It appears that he was also called upon to design and possibly direct court festivals, banquets, food decorations, and other ducal ceremonies.[15] He seems to have traveled a fair bit, much more than other artists of his circle. His sojourns included those with the duke himself, who often shuffled between his various residences in France and the Low Countries, as well as more distant travels, such as a diplomatic mission to the Iberian Peninsula in 1428–1429, whose aim was to secure for Philip the hand of Isabella of Portugal. (During this trip Van Eyck painted Isabella's portrait, possibly as a type of "photograph" for the Duke.) The truth of the matter would appear to be, in Craig Harbison's pithy phrase, that "Van Eyck was extensively involved in designing and making luxury objects,"[16] and the lushness of his materialism was as much or more a sign of the specific conditions of his daily life than of religious meditation. As a capable man of courtly ambition living where luxury and fanfare, but also piety, were greatly prized, Van Eyck seems to have painted the world of objects as a kind of territory of the eye, a realm of nearly unlimited visual possession and display. He may well have sought in his landscapes both refuge from and rehearsal of the decadent cosmopolitanism that defined so much of his existence.

Whatever interpretation we may want to place on his motives, they matter less than their result. Van Eyck's achievement, after all, must be understood as a kind of historical talisman. He reveals two crucial things: first, that the physical world was finally being redefined as a goal of observation and illustration; second, that the pictorial image in Western culture was about to evolve from a narrative and literary role—the telling of stories and the manifesting of the written word—to an epistemological one, where naturalism of portrayal could be a crucial means for inscribing, embodying, and conveying knowledge. Erwin Panofsky's well-known comment that "Jan Van Eyck's eye operates as a microscope and as a telescope at the same time"[17] should be considered in this light. Indeed, it can hardly be a complete coincidence that this eye and these instruments originated in the same locale of Europe. It is as if the painter had the power to offer us, *ante facto,* a vision of the transformational capabilities that human sight would one day come to have when extended by technology. Such is the depth of significance revealed by the image of the Moon, as drawn by this painter's hand.

This Moon, too, is an enormous distance from the textual arguments still raging at the time. These continued to deal with the legacy of Scholasticism and its concern with purity or impurity of the lunar substance, its rarity or density, its luminous properties, and so forth. It would be a century and a half before the next great transformation in astronomy would occur; this would quickly involve both written and drawn aspects of the heavens and, in Galileo's hands, would focus itself largely upon the Moon. Copernicus and Galileo were near contemporaries; they, too, were separated by the advent of the telescope, which placed them on either side of a vast observational divide. The truth is that the pattern for Galileo was already clear: the ability to see and record, as evidence of witness, had outpaced the interests of Ptolemaic astronomy. From Van Eyck onward, the Moon of Scholastic conception had thus begun to set.

LEONARDO DA VINCI AND THE RENAISSANCE

In his writings on the art of painting, Leonardo offered the following advice to all who might take up the brush:

> The mind of the painter must resemble a mirror, which always takes the colour of the object it reflects and is completely occupied by the images of as many objects as are in front of it. Therefore you must know, O painter! That you cannot be a good one if you are not the universal master of representing by your art every kind of form produced by nature. And this you will not know how to do if you do not see them, and retain them in your mind. Hence as you go through the fields, turn your attention to various objects, and in turn look now at this thing and now at that, collecting a store of diverse facts selected and chosen from those of less value.[18]

Elsewhere in his *Treatise on Painting,* Leonardo refers to art as a necessary pathway to science. This is because painting seeks to reproduce human sensory experience, which itself is "born of nature." Striking as these words are, we can see that Leonardo himself fell short of them in a number of areas, particularly when compared with Van Eyck. Yet they bring us to the endpoint of the naturalism begun centuries before, and they help direct us to understand that da Vinci also drew the Moon in a new way. He drew it as a self-contained "study," an isolated specimen placed upon its own page. His Moon was a kind of field sketch and thus represented the next step beyond Van Eyck.

Whether Leonardo produced more images of the Moon than the three of which we have spoken is not known. Certainly he wrote a great deal about the lunar orb and its nature, substance, and luminosity.[19] These writings, which are often musings on one specific subject, coalesce upon a central image, which is this: "[T]he part of the Moon that shines consists of water, which mirrors the body of the Sun and reflects the radiance it receives from it" (p. 162; also pp. 157–160). Moreover, only the crests and troughs of the waves in the lunar oceans reflect the sunlight, this being the reason that the light issuing from the Moon is so much weaker than that of the Sun (p. 160).

This bit of speculative reasoning is mixed with concrete deduction based on observation. Following Plutarch, Leonardo discounts the mirror theory of the Moon by virtue of the constancy of the lunar spots. He refutes Oresme's comparison of the lunar substance with alabaster, for example, and the more common analogy with crystal or glass by saying that, if such were the nature of the Moon, the spots would vary with the different phases, eclipses, and other changes that regularly occur. Similar reasons are given to discount the Scholastic idea of rarity and density making up the surface, Leonardo's refutation here being that during an eclipse, the Moon would not appear so uniformly dark, since "the solar rays would pierce through the portions which were thin" (p. 166). In slightly disguised form, this is the argument offered in Dante's *Divine Comedy* against the very same hypothesis (see chapter 5).

The most remarkable statements made on this general topic are those that propose the lunar body not as something unique and celestial, but as another Earth:

> If you keep the details of the spots of the Moon under observation you will often find great variation in them, and this I myself have proved by drawing them. And this is caused by the clouds that rise from the waters in the Moon, which come between the sun and those waters, and by their shadow deprive these waters of the sun's rays. Thus, those waters remain dark, not being able to reflect the solar body (p. 167).

Here is Pliny turned on his head. It is now from the lunar surface that moisture ascends to obscure our sight. Leonardo is loyal to the ancient belief that the Moon is fundamentally linked to ideas of wetness, yet in his description of the lunar oceans, he also brings us back to geography, in particular to the possibility of maritime voyages "beyond Thule."

"Lustrous waves," "clouds," "oceans," and "land" compose the lunar world. It is a concept that had its correlative in ancient times and that existed down through the Middle Ages. But Leonardo returns to it the concrete dimension it had in antiquity. He even wonders "how the spots . . . must have varied from what they formerly were, by reason of the course of its waters" (p. 167). Where there are seas and waves, there are currents; and where currents exist, there is the possibility of navigation. Written during the opening decades of the Age of Exploration, Leonardo's words, no less than Van Eyck's images, proved to be a harbinger of what was to come. It took such ultimately cosmopolitan men, so endemic to their own terrestrial moments, to help make the Moon a world that might be seen, drawn, and investigated. For the first time perhaps, the lunar surface seemed close enough to be far away: "[C]onstruct glasses," Leonardo says a century before Galileo, "to see the moon magnified" (p. 168).

7

The British Contribution

WILLIAM GILBERT AND THOMAS HARRIOT

FIRST MAP OF THE MOON:
WILLIAM GILBERT AND THE POLITICS OF NAMING

*I*t was in Britain that the Moon's history on Earth entered its modern phase. Sometime during the 1590s, the English physician and student of magnetism, William Gilbert (1540–1603), assembled a number of his cosmological ideas into written form and included with them what must be called the first true lunar map (fig. 7.1). This image, like Leonardo's, never appeared in its author's lifetime. Gilbert was known in his day (and is mainly recognized today) as the discoverer of terrestrial magnetism, which he described in a different work, *De magnete,* published in 1600. He died soon thereafter, and his papers were collected by his younger half brother under the title *De mundo nostro sublunari philosophia nova* (New philosophy of our sublunary Earth) and presented to Prince Henry, from whose hands they were circulated among other scientists of the day, such as Francis Bacon and Thomas Harriot, who therefore must have seen Gilbert's Moon image.[1]

Scholars are generally divided with regard to calling Gilbert's image a "map."[2] But a map it surely is, and a very interesting one. We can see from

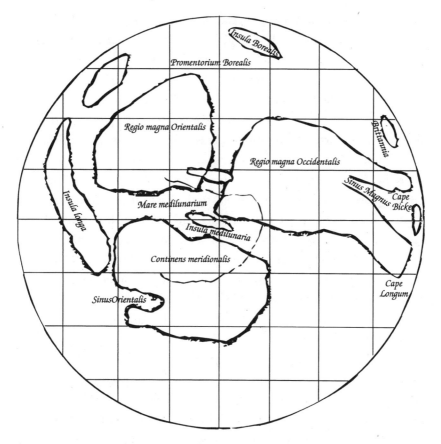

Figure 7.1. Naked-eye map of the Moon by William Gilbert, drawn about 1600. Grid lines exist on the original; however, that image is not suitable for reproduction. Adapted from Whitaker, "Selenography." Redrawn by Floyd Bardsley.

figure 7.1 that it is not an attempt at naturalistic portrayal—like the images of Van Eyck and Leonardo—but a flattened and abstract projection. Its greater interest to any cultural history of astronomy lies not merely in its visual aspects but also in its textual ones. Not satisfied with merely drawing the dark regions or "spots" of the Moon, Gilbert did something utterly remarkable: he named them, even portions of them, all after earthly geographic forms. Gilbert's concept of the planets, including the Moon, was a complex mixture of medieval, Renaissance, and modern (Copernican) attitudes. Like his scientific work in general, it displays the effects of metaphysical and Scholastic notions, tempered by a strong proclivity toward direct observation, experimentation, and the use of instruments.[3] To Gilbert, "The planets were solid, light-reflecting globes, closer to the Earth than the stars, not connected to any [Aristotelian] sphere, surrounded by their effluvia, and moving from the 'impulse' given them at the time of their creation."[4] There was much of medieval Neoplatonism in this. Gilbert noted that each planet had its own soul and unique quality. The Moon, for example, lay within the Earth's "orb of virtue," a term possibly equivalent to "magnetic field" yet more immediately derived from Plato's animistic views of the universe, re-stated by Albert the Great as "nobilities" for each planetary body. The lunar orb, moreover, acted as a "companion" to the Earth and could be described as a smaller version of the latter. Siding with those such as Plutarch and Leonardo, Gilbert saw a distribution of land and ocean on the lunar surface and proposed that this accounted for the variations in brightness. The Scho-lastic idea of opaque and transparent areas on the Moon he discounted en-tirely. Unlike Leonardo, he viewed the dark areas as land and the bright areas as lunar oceans—this made more sense within the paradigm that the Moon shone by reflected light alone and was not "obscured" by clouds.

It is thus with Gilbert that one of the most ancient beliefs about the lunar orb, a belief reaching back to the time of Orphic poetry in preclassical Greece (see chapter 3), becomes formalized at the level of the map, i.e., pictorial display and nomenclature. Gilbert was at least partly aware of this; his desire to impose a certain mimesis between Earth and Moon may have played its part in arguing that a naming scheme finally be applied to a surface gazed upon for so many centuries by human eyes. Gilbert actually bemoaned the fact that no one before him had done this, especially in antiquity, because the lack of older maps prevented him from discerning any changes that might have occurred in the Moon's face.

What kind of names were these first lunar titles, and how did they fare historically? Gilbert's choices were nearly all descriptive. Each title was meant to reveal the location of its referent, and thus the rationality for its choice. "Regio Magna Occidentalis" ("Great Western Continent"), for exam-ple, was chosen for one large portion of the lunar spots. Part of this continent was dubbed "Cape Longum" ("Long Cape"); another part, "Sinus Magnus" ("Great Bay"). On the upper side of the map, Gilbert perceived an island he titled simply "Insula Borealis" ("Northern Island"), and he called the great ocean separating the three continents "Mare Medilunarium" ("Middle Lunar

Sea"). Such an approach to naming a foreign terrain seems simple enough, even "primitive." Yet one should take note of an effect stemming from its reconnaissance style. By giving each feature a descriptive title, this approach creates "place" as a textual phenomenon. Geography is here a code of placement more than of observation per se.

In fact, in using such a system, Gilbert was mainly following an already established tradition in European mapmaking. This tradition had begun just over a century earlier with the first illustrated editions of Ptolemy's *Geographia.*[5] Ptolemy's text, compiled in A.D. 160, had been brought to Italy from Byzantium in 1406, translated into Latin shortly thereafter, and first published in the late 1470s with maps displaying more than eight thousand place-names. It is on these early maps that terms such as *sinus, regio, mare, insula, borealis,* and so on were established as generic titles; it is here as well that different portions of the Earth were given location descriptors: India intra, India extra, Java major, Java minor, Magnus Sinus, etc. By Gilbert's day, of course, mapping had developed enormously beyond the rudimentary versions drawn to accompany Ptolemy's text. Yet such Ptolemaic descriptors were still very much in use for various portions of the globe, especially those distant from Europe.

It seems significant that Ptolemy had an equal or greater influence over conceptions of the Moon through his writings on terrestrial geography. Geography indeed had superseded art in vying for the lunar surface; but not geography alone, and not as we conceive it today. The era of colonialism had raised the craft of mapmaking to a very high level of importance, and the status and distribution of maps as objects of documentation, study, and private possession had spread to every major capital of Europe. In Elizabethan England, too, maps were highly prized as works of curiosity, knowledge, art, and decoration.

Aiding this was a profusion of travel literature, which bloomed from the 1540s onward, culminating in Richard Hakluyt's eloquent and widely praised *Principall Navigations, Voiages, and Discoveries of the English Nation.* This work was first issued in 1589 and enlarged into three volumes in 1598–1600, was known in its day as the "epic of the English nation," and was one of the most widely read and influential texts during the sixteenth century. (Shakespeare drew some of his knowledge and scenery from it.) Hakluyt is credited with introducing a great deal of late Renaissance geography into

England, in part through his teaching during sixteen years at Oxford but more importantly through his writings and general propaganda for English settlement in the New World, especially North America. One of his major efforts in *Principall Navigations* was to make known the "fact" that the English had always been a great seafaring people, that the voyages of the Anglo-Saxons centuries before had been among the great accomplishments of the age, and that exploration and adventure were "in the bloode" of the English people.

Hakluyt was part of an important movement in late-sixteenth-century letters, sometimes called "antiquarian" but more accurately described as a movement to define, celebrate, and revitalize the endemic virtues of the English in the face of growing hostility from Catholic Europe (including the Pope). Spenser and Shakespeare were both part of this movement, though in different ways. Hakluyt, along with other influential writers on geography such as John Speed, saw England's survival and future greatness as being tied to expanding its people and territory, especially into North America, where Spain had not yet secured a firm hold. Spanish settlement (let us call it that) had proven the riches to be culled from exploration; such wealth was making Spain enormously powerful, able to impose its will upon other nations. On the plains of history, Spain was the great barrier to the English nation fulfilling its "true destiny." The truth of this destiny was soon proven in 1588 in the great battle with the Armada. English victory had been magnificent, breathtaking, foretold; but it also had to be secured if Spain was to be prevented from rising to its former glory. Hakluyt was instrumental in persuading Sir Walter Raleigh to devote his energy and resources to founding the first British colony in Virginia in 1584–1586. Often noted is that this name, "Virginia" (probably coined by Raleigh himself), was applied to the whole of the new territory and had a political-moral dimension, implying the superior virtue of a chaste Queen. The same, of course, was true of "James Towne," founded a few decades later (1607), though without the suggestive link of a "virgin territory."

All of this lay behind a single designation Gilbert chose for his map: "Britannia." Gilbert, we might note, was royal physician first to the British navy, then to Queen Elizabeth herself, and finally to James I. He no doubt admired and perhaps even knew Hakluyt as a writer. In the heady days of the 1580s as tensions gathered against Spain, it would have been difficult, if not impossible, to locate anyone close to the Crown who was unroused by the avid patriotism, vision, and ambition of this geographer magnus of the British nation. Moreover, it is evident that Gilbert himself had frequent contact with maps; his interest in magnetism led him directly into concerns with navigation, geography, astronomy, and cosmology. "It is significant with respect to the origin of Gilbert's interest in scientific accuracy" (one of the aspects of *De magnete* most often praised by contemporaries) "that all of his physical instruments are actually nautical instruments or are at least [based on them]."[6] He also possessed a large library and a collection of celestial and Earth globes. These, along with other possessions, were left to the Royal College of Physicians' library, tragically destroyed by the Great Fire of 1666.)

"Britannia" was an ancient Roman title, given to a land of distant and difficult conquest. Before the sixteenth century, the name "Britannia" was used in a historical sense only, but under Henry VIII and Edward VI, when efforts were made to annex Scotland, it came back into political usage as an emblem for English manifest destiny. William Camden's popular work, *Britannia, A Chorographicall Description of the Most Flourishing Kingdomes of England, Scotland, and Ireland, and the Ilands Adioyning, Out of the Depth of Antiquitie,* appeared in 1586 in the same turbulent decade as Hakluyt's book and equally promoted the concept of geography as fate.

It was this merging of political, military, geographical, and literary trends that Gilbert codified in using "Britannia" for one of the island portions of the Moon. England had beaten Spain; it would soon absorb the lands surrounding it and expand deeply into the New World. Did it not, therefore, have a deserved place on a new world of another kind, a world otherwise uncolonized, untouched, still in its "virgin" descriptive state? "Britannia" planted a claim on the Moon and implied that England, the new naval power of Europe, might one day send ships of a different kind to these distant seas and lands.

One other title of Gilbert deserves mention. The name "Cape Bicke," whose referent I have been unable to determine, nonetheless seems coined after a person or place of importance to the author. Like "Virginia," it signifies the power of the colonial discoverer to canonize those he favors in the form of a textual designation that is then transformed into universal, geographic space. This power, of course, had a still older vintage in European geography, beginning at least three hundred years earlier with the maritime trade empires of the Italian city-states and with Portuguese exploration of Africa. Gilbert's onomastic scheme for the Moon indicates that a blending of traditions had become attached to the idea of a "new world," traditions that included the classical, the contemporary-political, and the colonial. All of these traditions expressed the cultural sensibilities of the time. Such are the larger meanings inscribed upon the lunar surface by the first map.

Finally, we should note that Gilbert's basic reason for drawing the Moon was entirely different from that of Van Eyck or Leonardo. By his day, it had become common to include in scientific books many naturalistic drawings, artistic decorations, visual aids, and accents of all kinds, for both the instruction and the entertainment of the reader. Gilbert himself is exemplary in this regard; his *De magnete* contains a large number of excellent images meant to demonstrate his equipment, experiments, and conclusions. By this time, textuality was no longer sufficient; images now carried a weight of demonstration and evidence. Gilbert's map was itself a type of visual experiment, an attempt to demonstrate through inscription his conclusion that the bright areas of the Moon's surface were water, the dark areas land, and the whole a true territory that might one day belong to England.

GILBERT AND THE CELESTIAL ATLAS:
A BRIEF CORRESPONDENCE

In 1600, the same year *De magnete* was published, a series of gores (almond-shaped sections used to cover a globe) showing a map of the heavens was produced by Willem Janszoon Blaeu, a young astronomer who had worked with Tycho Brahe at his observatory. Blaeu was soon to become one of the most famous mapmakers of the seventeenth century and to gain further renown by being appointed official cartographer of the Dutch East India Company. His map of the skies was one of the first to depict the newly coined southern constellations, which the Dutch navigator Pieter Dircksz Keyser had delineated during a trip to Java. These stars had been used in navigation (and apparently named) by the Portuguese since nearly two centuries before this, but no maps or descriptions had ever been published, doubtless for reasons of secrecy. Keyser grouped the stars he observed into twelve figures, to which he gave the names of mythical beasts (Chameleon, Phoenix, etc.) or creatures native to the lands of southeast Asia (Bird of Paradise, Flying Fish, Toucan, etc.)—in effect, mixing a familiar pantheon with an exotic one. These names were firmly established when Johann Bayer published his magnificent celestial atlas, *Uranometria,* culminating nearly a century of star mapping and setting the standards for the genre thereafter, down to the nineteenth century.

Bayer's atlas of the constellations was a very different sort of publication than the literary astronomy found in printed versions of Hyginus and Aratus published in the late fifteenth and sixteenth centuries. The decorative images in these latter works qualify as a type of art stripped of science. Bayer instead followed a tradition begun, perhaps, by Albrecht Dürer, who produced an actual star chart in 1515 with a rudimentary grid system. More significant was Johann Schöner's early celestial globe (1533), followed by a more elaborate and famous example by Gerard Mercator in 1551. The most immediate predecessor to Bayer, however, was Giovanni Gallucci, whose *Theatrum mundi* (Theater of the world, 1588) adopted star positions from Copernicus's *De revolutionibus* and included elegant drawings of the constellations that were mapped using a trapezoidal system of projection meant to display equal-area sections. Gallucci's choice of titles was not without significance. It was taken, with only slight modification, from the grandest and most widely known terrestrial atlas in the sixteenth century: Abraham Ortelius's *Theatrum orbis terrarum,* first published in 1570.

What Bayer succeeded in perfecting, therefore, was not simply a way to draw and plot stars and star figures, but a style of visual conception. The heavens by this time were part of geography, inseparable from the earthly theater in terms of their representation in map form. Bayer included fifty-one star maps in his atlas, one for each of the forty-eight constellations in Ptolemy's *Almagest,* plus two planispheres and a chart of the southern skies. Each of his engraved plates contained a detailed grid that allowed star positions to be determined to less than one degree. He also introduced an augmented naming scheme by which letters of the Greek alphabet were used to classify individual stars on the basis of magnitude. Bayer's stellar reference system, which remains in use today, thus achieved something comparable to Gilbert's map of the Moon. It provided a descriptive code for identifying both the locations and the types of its referents. The title "Alpha Centauri," denoting the brightest star in the constellation Centaurus, bears a certain similarity to the name "Mare Medilunarium," designating a middle sea on the Moon. This type of reconnaissance vision is a geographical sensibility that astronomy has never fully abandoned because it would mean leaving behind the very idea of "exploration." Gilbert and Bayer were both observers and nomenclatural surveyors. The correspondence between their naming schemes, reflective of a renewed relation between astronomy and geography, is our sign that the heavens had now been transformed into "territory." Geography had left the Earth, only to return, more expansive than ever.

HARRIOT'S DRAWINGS:
THE MATHEMATICAL TRADITION

Gilbert's map of the lunar surface reveals that an interest combining Van Eyck's observation, Leonardo's isolating "study," and the explorer's desire to make territorial claims had evolved. The Moon was now in the hands of science, but also of geography, and therefore of politics. It seems possible that Gilbert had drawn the Moon with navigational benefit in mind, specifically as a first step toward solving the problem of determining longitude (one of the great practical obstacles of the age). Such a purpose would have suited very well his position as royal physician to the British navy and his standing as a recognized luminary of British science. The data, however, are lacking; the author speaks only of his desire to correct the errors of the past and substitute a truth of the present. His Moon remains descriptive, and his names are correlative with the earliest phase of colonial exploration.

Gilbert's effort at lunar mapmaking, however, can also be read as a sign that the Moon's surface would soon enter the main currents of astronomical illustration. What brought this about was the invention of the telescope. This instrument, newly available in Holland, fell into the hands of one of Gilbert's younger contemporaries, the English mathematician, cartographer, and astronomer, Thomas Harriot (1560–1621), who proceeded to observe and draw the lunar surface in a manner that effectively married the tradition of mathematical diagrams (minus any allegorical figures) with naturalistic portrayal. Harriot was well known as one of the foremost mathematicians of

Elizabethan England and enjoyed a significant reputation among the British scientific establishment as the equal or better of any thinker on the continent.[7] Like other brilliant men of his day, Harriot had his hand in many areas of intellectual endeavor and is well known as the close friend, teacher, and technical adviser of Sir Walter Raleigh. Inspired by Hakluyt, Raleigh had his own designs upon the New World and employed the young Harriot to instruct him in cosmology and navigation, then sent him to Virginia in 1585 on an early reconnaissance voyage. The intent of this voyage was to make an inventory of the land and people in the hope of providing information helpful for future colonial settlement. Together with the artist and mapmaker John White, Harriot planned to produce an extensive, highly detailed account of everything seen on his visit, a kind of experience-based encyclopedia of the New World. A terrible hurricane cut their mission short. Because of a hasty departure, all that could be salvaged were Harriot's *A Briefe and True Report of the New Found Land of Virginia,* published in 1588 in roughly thirty pages, and a scattering of White's magnificent drawings and maps, the former of which depict in remarkable detail the native inhabitants, flowers, fish, crabs, turtles, insects, and a host of other flora and fauna.[8]

Although Harriot would spend a lifetime investigating advanced problems in optics, physics, astronomy, and above all, mathematics and would make significant contributions to each of these fields, his *Briefe and True Report* remained the only publication attached to his name during his lifetime. Harriot appears to have been a reserved man, a "perfectionist" unwilling to publicize his research, which was extensive in several branches of mathematics.[9] In his *Briefe and True Report,* Harriot is descriptive, concise, and trim in his choice of language and facts. He took care to learn something of the Algonquin language and to use native Indian names for everything that was foreign to him—indeed, a fair portion of his book reads like a botanical glossary. *Briefe and True Report* went through many editions in only a few years and was translated into Latin, German, and French, in addition to being later reprinted in volume three of Hakluyt's own *Principall Navigations* (1598–1600). Its popularity reveals the public interest in tales of British travelers. In fact, Harriot's account of his visit begins in a prophetic way with regard to his later "voyage" to a different type of New World: "I have therefore thought it good, being one that have been in the discovery . . . to impart so much unto you [that] by the view hereof you may learn what the country is, and thereupon consider how your dealing therein may return you profit and gain."[10]

It is interesting to think of what might have resulted had Harriot, then in his twenties, viewed the Moon through his telescope instead of two decades later. Rich with visual experience, with the conjunction between maps and actual territories, and with the power of imagery and words to capture the "distant" and "exotic," what might he (and possibly White) have made out of the Moon, seen in magnified form? We will never know, of course. Following this expedition, Harriot retreated into more theoretical work on mathematical navigation. During the 1590s, we find him working with Hakluyt on the technical construction of maps, developing new methods for determining and plotting position via techniques he himself had devised.[11] From this point on, Harriot acted as a technical advisor only, and the world of his visual use was mainly restricted to flat mathematical diagrams.

In 1609 Thomas Harriot became one of the first learned men in England to get hold of a new invention, the telescope, which had emerged from Holland during the previous fall and was circulating among European royalty and being imitated by spectacle makers in the Spanish Netherlands, England, and elsewhere.[12] Harriot's own interest in this new instrument was instantaneous, since he had been working on various problems in optics for some years. On July 26, 1609, he became the first person to produce an image of the Moon on the basis of telescopic observations, preceding Galileo by only four months. Harriot apparently used a six-power telescope, and he drew a five-day-old Moon.[13] As shown in figure 7.2, this image is a rather crude rendering of the crescent phase, with an irregular terminator (shadow line) and several maria roughly darkened (Mare Crisium, Mare Fecunditatis, the eastern portion of Mare Tranquillitatis, and Mare Nectaris). It is more a sketch than a drawing, and it is completely without names or designations— a lack that would argue for the immediacy of the image. Moreover, it is accompanied by no commentary or written observations. It would appear that after jotting it down, Harriot did not consider this drawing to be very important: he did not spend much time on it, did not expand it textually, and did not draw any other such images in the days, weeks, or even months immediately following. The instrument may well have been of poor quality,

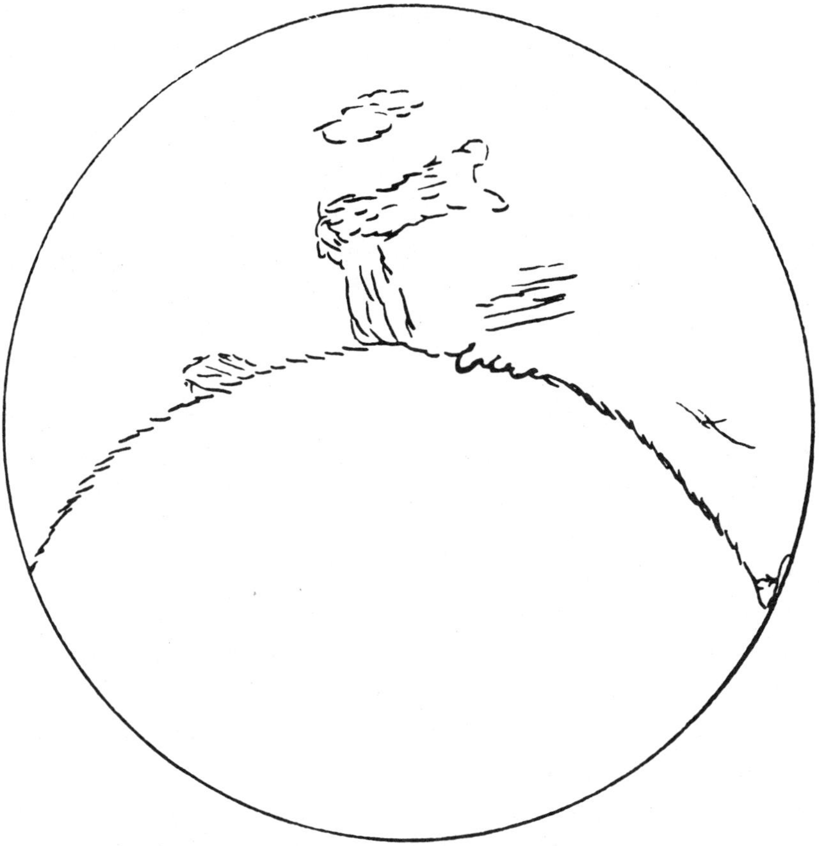

Figure 7.2. First known telescopic drawing of the lunar surface, dated July 26, 1609, by Thomas Harriot using a six-power instrument. Courtesy of Ewen Whitaker.

but even if it had been as good as or better than the telescope used by Galileo, it seems unlikely that Harriot would have produced a more naturalistic or even cartographic version.

Yet this drawing is extremely significant and not simply because of its historical priority. It shows that the telescope, as an instrument of expanded sight, had the power to outdo the textual imagination. For Harriot, the first impulse of discovery was to record the seen in the form of a picture, not to translate this vision into words. In his mission to Virginia, he had been guided by the demand to touch, taste, interrogate, and describe the phenomena before him, to carefully enter the objects of the New World into a catalogue of names and textual identities. This time, gazing into his new instrument, Harriot could not escape the event of sudden and distant witness that seems to have consumed any documentary impulse.

The deeper problem for Harriot, however, was that he simply was not equal to the task. Here was the chance for Van Eyck's eye to prove its legacy, for Leonardo's hand to complete its work. Yet the talent went lacking. It had to wait several months for another embodiment. A year later, after having received one of the first copies in England of Galileo's *Sidereus nuncius* (Messenger from the stars, published in March 1610), complete with remarkable drawings of the lunar surface, Harriot was inspired to produce new images himself, this time with actual topographic features, obviously in direct imitation of his Italian colleague.[14]

It is evident that Harriot's interest was rekindled by Galileo, for he made no fewer than four drawings of the Moon in July (1610), seven in August, and nine in September.[15] The drawings made in July, like that of the year before, have no commentary associated with them. But by August, observations finally begin to appear. On August 21, for example, Harriot writes, "nothing notable"; on the 22nd, he observes "inaequalities scarce sensible . . . a little ragged with a peninsula"; on the 26th, "some partes having greater eminences & some valleyes with shadowes." Then on September 11 we learn, "The appearance was notable, ragged in many places . . . with some ilands and promontoryes"; and by October 23, "it shewed mountenous . . . that also was montaynous with an opening in the middest & some black passage from it."[16]

What one sees here are words catching up with a perception in search of a discourse. The progress of Harriot's commentary reveals him as he discovers an ability to see the Moon directly in terms of new textual and visual inscriptions, both supplied by Galileo. Harriot seems to discover this terrestrial Moon in several stages, beginning with oceanic forms and progressing to land proper in his list of images—seas, islands, peninsulas, valleys, and mountains, all jostling each other in dense proximity. Harriot named only one feature, "The Caspian," which apparently corresponded to the Mare Crisium, the most isolated, i.e., "land-locked," of the lunar maria.[17] Harriot, along with Francis Bacon, was one of the few men who had seen Gilbert's *De mundo* in its earliest form, possibly even before 1609,[18] and thus may have been ready to "see" a literalized version of terrestrial forms on the Moon. This was no mean feat, however, and Harriot's apparent struggle to do so, after much time, study, and even outside example, is a sign that comprehending the Moon in this manner meant a local change in the historical capacity of human perception.

We have other evidence of this. Letters written to Harriot by his one-time student and later scientific colleague, Sir William Lower, reveal a fascinating before-and-after account of "seeing" the Moon in the historical umbra of Galileo. Harriot sent his first telescope to Lower in late 1609 or early 1610, advising him to pursue his own observations. In a letter dated February 6, 1610, Lower wrote back to thank Harriot for "the perspective cylinder":

> According as you wished I have observed the Moone in all his changes . . . [Near] the brimme of the gibbous part towards the upper corner appeare luminous parts like starres, much brighter than the rest, and the whole brimme along lookes like unto the description of coasts, in the dutch bookes of voyages. In the full she appeares like a tarte that my cooke made me the last weeke. Here a vaine of bright stuff, and there of darke, and so confusedlie al over.[19]

Lower is groping here, seriously and playfully, to find an apt description. Words fail him; Harriot's own drawing (which also had been sent) provides little help. Stars, coasts, a tart, a confusion of light and dark: Lower is trying to make sense of what he sees and can only produce a surplus of images, a narrative "confusedlie al over." He descends from heaven to Earth to the breakfast table and finally disperses his attention into a state of being overwhelmed. The key lies within the very images he has invoked, the most ancient of all—why doesn't he "see" it? The Moon as a territory of land and sea rises here, like a phoenix, yet remains invisible. Moreover, Holland, the very origin of the European telescope, is called upon for a moment only to collapse into humorous dismissal. No discourse yet exists for what is seen, and like Harriot, Lower is not the one to create it.

In late spring or early summer of that same year (1610), Harriot wrote to tell Lower of Galileo's discoveries, and on June 11, Lower wrote back:

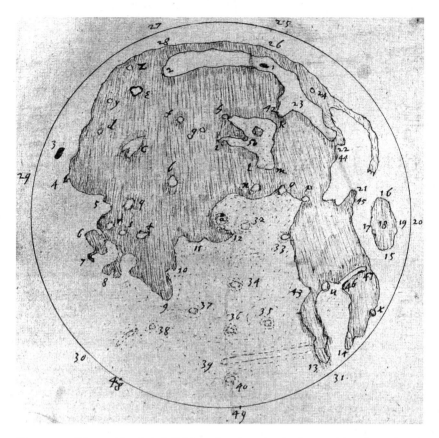

Figure 7.3. Thomas Harriot's full-Moon drawing, dated 1611 and based on telescopic observations. Courtesy of Ewen Whitaker. Reprinted by permission of the Earl of Egremont and Leconfield.

> Me thinkes my diligent Galileus hath done more in his threefold discoverie than Magellane in openinge the streights to the South Sea or the dutchmen that were eaten by the beares in Nova Zembla. I am sure with more ease and safetie to him selfe & more pleasure to mee. I am so affected with his newes as I wish sommer were past that I mighte observe the phenomenes also. In the Moone I had formerlie observed a strange spottednesse al over, but had no conceite that anie parte thereof mighte be shadowes.[20]

With a grasp of shadows and a few guiding words, the Moon for Lower is transformed into a rugged territory of terrestrial substance. The geographic metaphor returns: Galileo's eye has sailed as far as Magellan's and that of the Dutch. Calling Galileo a type of Magellan expresses Lower's "seeing" the Moon as another Earth and helps him preserve this new perception by giving it a meaning entirely contemporary to the time.

In the year after this exchange, Harriot produced yet another image, the first drawing of the full Moon as seen telescopically (fig. 7.3). This image combines an attempt to outline the dark regions of the lunar surface with the use of letters and numbers whose only purpose is to help correctly position these regions, not to designate actual features. It is thus only partly a map—similar to Gilbert's delineation of coastlines—being part diagram as well. Because it shows none of the topographic features (e.g., craters) that Harriot had drawn the year before after seeing Galileo's book, it seems regressive, pictorially speaking. The Moon is returned to being a massive object, divided only into light and dark as it had been for thousands of years before. Harriot's failure seems one of comfort. Perhaps conscious of his imitative efforts after Galileo, he tried for something different, a full-Moon image drawn in the manner he knew best—as a mathematical diagram. Even his pictorial images of the year before are surrounded and encased in a textual sea of few words but many computations, formulas, and calculations.[21] Harriot was happiest with the Moon as a bearer of mathematical opportunities and visibilities. Galileo, with his own magnificent perception, helped him see the lunar surface geographically, but this is not what had the most important meaning for Harriot. As he no doubt understood, borrowed vision is not the same as discovery.

8

Galileo

MAPS WITHOUT NAMES

AN EYE OF FAITH

The change that Galileo produced, by which the Moon finally became a fulfillment of Plutarch's prophecy, could only occur when the lunar surface was observed to have earthly features—when it was literally seen and convincingly represented as another Earth. The task was not one of perception alone, but of picturing, both in words and images. The Moon did not require features that merely looked like terrestrial ones: at some point, the lunar surface had to be shown to contain actual mountains, valleys, islands, peninsulas, seas, bays, and plains. This demanded not only interpretation but also extrapolation from rather meager information. As Roger Ariew has pointedly written:

> But what were Galileo's discoveries, his facts? Did Galileo discover mountains on the Moon? We assume that he did so because we know that there are mountains on the Moon. But what Galileo discovered, using his telescope, was that spots on the lunar surface change shape

over time; he then concluded that this phenomenon is best ex-
plained as the shadows cast by mountains on the Moon if the light of
the sun is reflected off the Moon.[1]

The telescope, then, was not the absolute determining factor we often
conceive it to have been. At the same time, Ariew's claims, I believe, are a bit
exaggerated. Galileo saw a great deal more than mere changes in the shape of
lunar spots. As he makes clear in *Sidereus nuncius,* he observed the irregular
advance of night across the lunar surface, noting salients and entrants of
light, emerging points in seas of shadow that could only indicate an uneven,
complex surface. Kepler had made the same conclusion some years before in
his book on optics (*Astronomiae pars optica,* 1604) from the irregular edge of
the terminator.[2] Nonetheless, Ariew's larger point is undeniable: Galileo did
not *see* mountains and valleys on the Moon. He saw other things—a jagged
outer edge, evolving and irregular patterns of light and shadow—and the
meaning he gave them called upon a certain eye of faith in a non-Aristotelian
lunar truth.

At this point in the history of the Moon, the borders between analogy
and visual reality dissolved or, as suggested by Gilbert's map, no longer
existed. Galileo was more than aware of Averroes's ideas regarding a homogeneous lunar substance subject to "rarity" and "density"; he may even have
subscribed to such ideas in his youth. It seems likely, however, that he had
changed his mind before gazing through the instrument that was to change
the heavens forever. It may well have been Plutarch's work *The Face in the
Moon* that altered Galileo's view. It may also have been this text in conjunction with Kepler's work on optics, which speaks a fair bit about the nature of
the lunar surface, its mountains, its atmosphere, and so on.[3] The Moon-as-
another-Earth idea was a very ancient one by the time Galileo adopted it—
one of the most ancient of all—and it had gained many vocal adherents by the
sixteenth century.[4]

The story of Galileo's telescopic observations has been told in detail too
many times to be repeated here.[5] If I have spoken more of Gilbert and
Harriot in this regard, it is because they have always been assigned too
confined and minor a niche in the scholarly edifice of the history of astronomy. As we have seen, Harriot reveals that an eye and hand inadequately

trained for the task of artistic documentation proved the pinch in the historical hourglass. If art preceded science in this area, as Van Eyck shows it did, then art was needed again to make the new vision real. Art seems to have caught up with science at last in Galileo's drawings, which deserve all the attention they have received and more.

Galileo built his own telescope, eventually producing an instrument capable of twenty-power magnification (much more powerful than Harriot's), which he turned toward the four-day-old Moon on November 30, 1609. In *Sidereus nuncius,* he tells us exactly how he learned of the telescope, how he fashioned one himself, and then, step by step, what he did with it, where he pointed it, and what he saw. Galileo thus leads us through his experience, laying out the voyage of his discoveries. Harriot, who had actually journeyed to a New World and produced a report about what he saw and learned, did no such thing with regard to his observations of the Moon. Galileo's little book, however, offers a traveler's account: "In this short treatise I propose great things for inspection and contemplation by every explorer of Nature," the book begins.[6] What follows is a log combining text and image in a manner that is documentary as well as suggestive, even impressionistic. Where, then, does the eye begin its journey? "It would be entirely superfluous to enumerate . . . the advantages of [the telescope] on land and at sea. But having dismissed Earthly things, I applied myself to explorations of the heavens. And first I looked at the Moon from so close that it was scarcely two terrestrial diameters distant" (p. 38).

Galileo, the explorer, does not really abandon Earth for the heavens. Geography is the first thing that his narrative evokes, and it is the Earth that measures and defines the closeness of the Moon. Upon arriving at the lunar surface, Galileo mentions the "darkish and rather large spots" that "every age has seen," but he then goes on to point out "other spots, smaller in size and occurring with such frequency that they besprinkle the entire lunar surface"; these features, he emphatically notes, have been "observed by no one before us" (p. 40). The author is entirely conscious of his position as a discoverer, someone who sees for the first time what others have not seen. In naming himself the "messenger from the stars" or his book a "message from the stars" (both readings of the title are possible), he is being coy, humbly claiming to be only the receiver and deliverer of this announcement. Yet he is also very much its author and sender, one who has returned from a distant place, ready to "unfold great and very wonderful sights . . . to the gaze of everyone."

> These sights, though unknown, are not entirely unfamiliar however. It is most beautiful and pleasing to the eye to look upon the lunar body . . . from so near . . . Anyone will then understand with the certainty of the senses that the Moon is by no means endowed with a smooth and polished surface, but is rough and uneven and, just as the face of the Earth itself, crowded everywhere with vast prominences, deep chasms, and convolutions. (p. 36)

It is, therefore, in every way, the Earth that Galileo discovers. Wrong in one sense, those who viewed the Moon as a mirror were right in another. This becomes far more striking later on in the work, as we will see.

Between November 30, 1609, and January 9, 1610, Galileo produced a large number of drawings (eleven of which have been preserved) and a series of vivid descriptions of what he saw. Together these expressions solidified the Moon into a formidable geographic entity of magnificent detail. Galileo's great deed was not so much to create a new "pictorial rhetoric" for the Moon,[7] but to marry once and for all the convincing powers of artistic naturalism to the long-term preference in astronomical science for the written text. His little book, produced quickly and impressionistically, is a landmark in Western intellectual history for this reason alone. Such a marriage between art and science Galileo performed on more than one level. Many of his written descriptions are best thought of as verbal paintings; this is especially true of his description of the lunar sunrise. His text is accompanied by four engraved drawings showing the crescent, first quarter, waning gibbous, and last quarter phases only (no full-Moon images). It is evident that he recognized the "essence" of the Moon in the terminator shadow—this essence being no longer directly involved with the question of luminescence, but instead with the sight of landforms coming into view across the dawning surface (fig. 8.1). The engraved images, however, were not enough to demonstrate this character without the following textual portrait:

> [All of] these small spots just mentioned always agree in this, that they have a dark part on the side toward the Sun while on the side

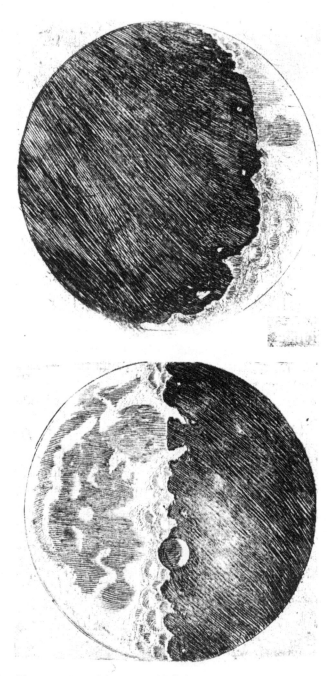

Figure 8.1. Two engravings of the Moon published in Galileo's Sidereus nuncius *(1610), based on ink-wash drawings by the author. Courtesy of Ewen Whitaker.*

opposite the Sun they are crowned with brighter borders like shining ridges. And we have an almost entirely similar sight on Earth, around sunrise, when the valleys are not yet bathed in light but the surrounding mountains facing the Sun are already seen shining with light. And just as the shadows of the earthly valleys are diminished as the Sun climbs higher, so those lunar spots lose their darkness as the luminous part grows. Not only are the boundaries between light and dark on the Moon perceived to be uneven and sinuous, but, what causes even greater wonder, is that very many bright points appear within the dark part of the Moon, entirely separated and removed from the illuminated region . . . Gradually, after a small period of time, these are increased in size and brightness . . . Now, on Earth, before sunrise, aren't the peaks of the highest mountains illuminated by the Sun's rays while shadows still cover the plain? Doesn't light grow, after a little while, until the middle and larger parts of the same mountains are illuminated, and finally, when the Sun has risen, aren't the illuminations of plains and hills joined together? (pp. 41–42)

Here, finally, are the "mountains flaming bright" that Plutarch had written of so long before. Here is the Moon brought fully within the gravity of terrestrial discovery and exegesis. Galileo writes to infect his reader with his own enthusiasm, mimicking the very image of a landscape emerging from darkness. He does this repeatedly, as the following passage reveals:

I would by no means be silent about something deserving notice, observed by me while the Moon was rushing toward first quadrature, the appearance of which is also shown in the above figure [fig. 8.1]. For toward the lower horn, a vast gulf projected into the brighter part. As I observed this for a long time, I saw it very dark. Finally, after about 2 hours, a bit below the middle of this cavity a certain bright peak began to rise and, gradually growing, it assumed a triangular shape, still entirely separated from the bright face. Presently three other small points began to shine around it until, as the Moon was about to set, this enlarged triangular shape, now made larger, joined together with the rest of the bright part, and like a huge promontory, surrounded by the three bright peaks already mentioned, it broke out into the dark gulf. (pp. 42–43)

As the first man on the Moon, Galileo is prone to an excitement of witness that belongs as much to travel literature as to astronomy. One can only imagine the overwhelming admiration such discovery must have elicited from Thomas Harriot and the rest of the European astronomical community—including, in generous if qualified fashion, one Johannes Kepler.

KEPLER: AN ELLIPTICAL ORBIT

Galileo had sent a copy of his book across the Alps to the royal mathematician of the Hapsburg empire in Prague, seeking Johannes Kepler's direct endorsement. He had done this via the Tuscan ambassador, who, after arriving in Prague, was leaving again in eleven days. During this brief period, Kepler composed a reply in the form of an open letter, entitled *Dissertatio cum nuncio siderio* (A conversation with Galileo's sidereal messenger), in which he stated his faith in the veracity of Galileo's findings and his commitment to do battle on the latter's behalf against "obstinate critics of innovation, for whom anything unfamiliar is unbelievable, for whom anything outside the traditional boundaries of Aristotelian narrow-mindedness is wicked and abominable."[8] Kepler had no telescope as yet and so could not verify the discoveries himself. Yet the "Imperial Mathematicus" (whose own works had been sent repeatedly to Galileo, with only minor response) also felt it necessary to point out that such an instrument had been envisioned a half century before, in Giovanni Battista della Porta's influential work, *Natural Magic* (1558). He then discusses a design that he himself intends to use that is superior to Galileo's instrument. A large part of Kepler's reply, in fact, is clearly meant to put a frame of precedence around some of his colleague's painted claims for originality. "These very acute observations of yours," Kepler writes, "do not lack the support of even my own testimony. For on page 248 of my 'Optics' you have the half-Moon divided by a wavy line. From this fact I deduced peaks and depressions in the body of the Moon. On page 250 I describe the Moon during an eclipse as looking like torn flesh or broken wood, with bright streaks penetrating into the region in shadow" (p. 27).

Kepler reminds Galileo that the ancients themselves had provided evidence for mountains on the Moon and that the relevant ideas had been justly and extensively recorded in Plutarch (a fact, he implies, Galileo knew quite well). Kepler says that, while still a student, he had used the methods of the ancients to estimate the height of some features on the lunar surface by measuring the length of shadows. "Therefore, Galileo," he sums up, "you will not envy our predecessors their due praise. What you report as having been quite recently observed by your own eyes, they predicted, long before you, as necessarily so. Nevertheless, you will have your own fame" (p. 38).

This is not where things end, however. Kepler informs Galileo that he (Kepler) has broadened his own earlier efforts on the lunar orb to produce "a complete geography of the Moon." With a characteristic mixture of humor and seriousness, he writes:

> As soon as somebody demonstrates the art of flying, settlers from our species of man will not be lacking. Who would once have thought that the crossing of the wide ocean was calmer and safer than of the narrow Adriatic Sea, Baltic Sea, or English Channel? Given ships or sails adapted to the breezes of heaven, there will be those who will not shrink from even that vast expanse. . . . Who would have thought that navigation across the vast ocean is less dangerous and quieter than in the narrow, threatening gulfs of the Adriatic, or the Baltic, or the British straits? Let us create vessels and sails adjusted to the heavenly ether, [for] there will be plenty of people unafraid of the empty wastes. In the meantime, we shall prepare, for the brave sky-travelers, maps of the celestial bodies—I shall do it for the Moon, you Galileo, for Jupiter. (p. 39)

Kepler plants his own claim on the Moon in the midst of declaring it the colonial territory of the future. In his enthusiasm, he is the boldest of spokespeople for his age because he holds up the mirror to "discovery" as a struggle among explorers, nations, and thinkers fully conscious of their historical place. With remarkable prediction, Kepler foresees—in an instant of mingled hope, anxiety, and perhaps envy—that spaceships with names like

Mariner, Pioneer, Voyager, and (yes) Galileo will one day be sent out to explore the solar system, bringing with them the entire tradition of colonial demand for knowledge and the riches of exotic territories. "The revered mysteries of sacred history are not a laughing matter for me," Kepler tells us (p. 40), and we should no doubt believe him.

Kepler's own lunar geography will be examined in detail in the next chapter. It is enough to say here that Kepler had been accumulating his ideas on the lunar surface ever since he was a student at Tübingen in the 1590s, when he had been struck with an aesthetic vision: how might the heavens look to an inhabitant of the Moon, who perceives his own world as immobile?[9] Kepler, however, like Harriot, was unequal to his vision in one major respect: he gave it no pictures, no visual language of witness. He produced no portraits or maps of the Moon, none of the evidence demanded by the period in which he lived. The closest he came is a rough sketch of a darkened portion of the Moon during an eclipse, with the terminator crudely drawn across the upper right. This image appeared in *Astronomiae pars optica* (1604) and bears a strong similarity to Harriot's own first drawing (see fig. 7.2), with its mathematical aspects. To put things bluntly, Kepler's own "complete geography" of the Moon was no geography at all in the great age of the map. His written notes, moreover, would be gathered and published only after his death, as a fictional journey to the Moon entitled *Somnium* (The dream). This literary effort was Kepler's own elliptical orbit around an original ambition to give the Moon a terrestrial meaning. In 1629 near the end of his life, despairing of political events, he wrote to his friend, the humanist Mattias Bernegger, "[S]ince we shall be driven out of the Earth, the book will be a useful provision for our journey to the Moon."[10]

Kepler was not the man for the Moon. As a literary author, he could not complete his book; even as astronomy leaped ahead by the grace of his pen, *Somnium* lapsed. As a writer, Kepler never sought to go beyond Galileo, to explore further the lunar surface and stake a true precedence. He never used the telescope to enter the fray of lunar image making, which became both intensely active and largely unproductive in the decade after *Sidereus nuncius* appeared. Geography, the map, the telling of an explorer's witness—these powers of the eye and hand he left to others. As a result, the rise of the Moon of modern science will forever be associated with Galileo.

GALILEO'S IMAGES: ART, SCIENCE, AND HISTORY

What of Galileo's engravings? What type of language did they give to the Moon? Beyond question, they are impressive documents (fig. 8.1). Their historical role was to help shift astronomy from a science with diagrams and allegories to one with pictures, whose later forms would include photography in the nineteenth century and digital imagery a hundred years later. *Sidereus nuncius* gave art the powers of evidence in astronomical science.

There are limits, however, to Galileo's new visual language for the Moon. We see clearly in Galileo's images the irregular terminator, with its "horns" and "islands" protruding into the shadowed portion of the surface. We see the maria complexly outlined, as well as a number of the craters (what Galileo termed "cavities") beautifully sketched along the left side of the terminator, their darkened portions on the sunlit side, just as described in the text, giving the surface a sense of deep and striking relief. We perceive all this and more. We do not, however, see mountains and valleys. Everything that might indicate a richness of topography has been stylized or stylistically exaggerated. We find, for example, by looking at a lunar photograph of the same general size, that Galileo has altered many aspects for effect.[11] The terminator is far more irregular than in reality, and the craters are enlarged to almost double their size. In the case of the largest crater, Galileo so magnified the original feature that it can no longer be identified with any certainty. (Presumably it is Albategnius, but there is no final agreement on this.) Little surprise, then, that his estimates of the height of lunar mountains, based on geometric analysis, yielded a figure more than four times that of any mountain on Earth.[12] Similar calculations had been performed in antiquity on the basis of naked-eye measurements.[13] We might recall (see chapter 3) that Philolaus had spoken of "living creatures . . . fifteen times as powerful as those on Earth." Galileo's lunar geography cannot be so easily divorced from such exaggerations. It seems at least ironic, as Ewen Whitaker says, that the Tuscan felt he had to portray the Moon in such fashion if he was to defeat a different classical conception, that of Aristotle.[14]

Early in his career Galileo had taken lessons in perspective from the mathematician Ostilio Ricci, which he had done in Florence along with the artist Lodovico Cigoli, a lifelong friend.[15] Galileo's first biographer, Vincenzo

Viviani, has also said that Galileo's original ambition was to become a painter; his father, however, discouraged such a career.[16] That he was knowledgeable in artistic technique is proven by the ink-wash images he gave to the engraver for the final pictures in *Sidereus nuncius*. These washes show he could create excellent, three-dimensional portraits of the Moon, not only of the planet as a whole but also of the individual craters that so strikingly compose its surface. The techniques Galileo employed were fairly simple; they involved shading above all. This perfectly suited his discovery that many of the small dark patches on the lunar face were caused by shadows. He thus did for the lunar form what Van Eyck and Masaccio achieved for the human figure: he gave it the fullness and solidity to cast shade. This, however, raises another question.

What models might Galileo have drawn upon for his rendering of lunar features? Perspective, shading—such are background skills. Galileo had to produce a new world in close-up view, and no artist could help him with this task. Where, then, did his visual ideas come from? As an inheritor of Renaissance art, Galileo the scientist would also have been heir to Renaissance cartography, the plotting and drawing of features on maps of the Earth and its various parts. No man of science in the sixteenth and seventeenth centuries could ignore the map as a vast new canvas on which enormous amounts of previously unknown information merged with artistic, textual, architectural, and mathematical aspects. Maps—not paintings—were the great documents of the age by which European civilization inscribed itself upon the world. We need to reconsider in these new terms the idea that Galileo perceived the resemblance of the lunar surface to the Earth.

There are two problems here. First, as any glance through a telescope will show, the Moon hardly looks like the Earth. Its bleached or blackened surface, pocked with circular craters and corrugations, mottled throughout, resembles no terrestrial phenomena that Galileo would have ever seen.[17] Second, no telescopic views of the Earth existed as a standard for comparison. The only thing that approached such a view was a map or globe, in any case an artificial projection. What Galileo had to "perceive," therefore, was an idea of the Moon visualized. His sight was conceptual as much as perceptual, but in a particular direction. In placing the Moon on a surface, he had to invert the geometry of witness—from looking up into the sky through a narrow "cannon" (as he called it, thereby implying its power to fire the eye outward into space) to a downward gaze at a projected image on a page. As suggested by the confusions of Thomas Harriot and William Lower, a photographic type of realism could not have done this. Galileo does not write about whether the lunar surface is rocky, watery, or whatever; he is not concerned with its composition, only with its gross geographic forms, the very forms that were placed on maps.

The distortions and exaggerations he visited upon the lunar face were more extensive than already mentioned. Aside from the terminator, to which he gave an excessively scalloped appearance, or the apocryphal "largest cavity," he also provided an overly smooth look to lighter areas within the western maria. The prominent "explosive" craters Tycho, Copernicus, and Kepler are missing entirely despite the fact that one or more of them would certainly have been plainly visible, particularly given the number of other craters that Galileo drew. Galileo chose to edit the lunar surface so that it would look more Earth-like than it was, removing the most alien features and contouring others in accord with certain conventions of geographic representation for maps. Yet, as his text makes plain, we as readers are supposed to view the naturalism of his drawings as total or near absolute; each image stands alone, uninvaded by text of any sort, a self-contained embodiment of observation. This makes them very much like paintings and radically distinguishes them from Harriot's images, with their armor of computations and formulas. For Galileo, the image must convey its own language, apart from words; it must provide a convincing simulation of what was seen in the lens. Looking at these pictures, we are shifted across the boundary between "observation" and "artifice," between the Moon as seen via the telescope and the Moon as transported to Earth via art and the map. These images are far more than "visual aids"; they are attempted fixatives of sensibility, perception, and belief.

I am suggesting that Galileo drew the Moon according to certain conventions of pictorial rhetoric in late Renaissance mapmaking that governed the delineation of coastlines, islands, peninsulas, headlands, basins, and so forth. These conventions were guides that helped him sort through the mass of complex visual impressions. It is also likely that the engraver had a role in this, being perhaps experienced in the production of maps. Several conventions in particular appear to have been helpful. One of these was the tendency to draw known coastlines—those that had actually been charted by sight—in

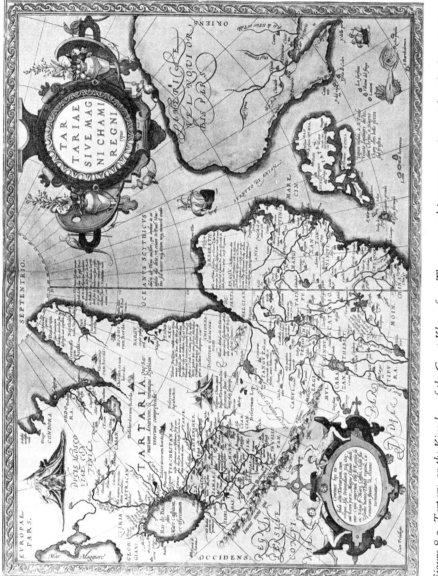

Figure 8.2. Tartary, or the Kingdom of the Great Khan, from Theatrum orbis terrarum *(1570) by Abraham Ortelius. Courtesy of the National Maritime Museum, London.*

a heavily scalloped fashion, often with armlike protrusions reaching out around bays or inlets (fig. 8.2). Unexplored or poorly known coasts and islands, however, were commonly given smoother edges, designating their reconnaissance status. Edges of coastlines were more darkly shaded than the open oceans to give the appearance of added relief and contrast (fig. 8.2). In the most famous atlases of the day, such as those by Ortelius (1570s and 1580s), Mercator (1580s and 1590s), and Hondius (before 1610)—all of whom were Dutch by birth—lakes were also darkened around their edges, giving them a subsided, craterlike appearance. Mountains, too, had been added by the mid- to late sixteenth century, but were depicted crudely, as long rows of bumpy masses having an emblematic function rather than a descriptive one. Shading to give some character of three dimensionality was done either on the western or eastern side to suggest shadowing at sunrise or sunset (the eastern or sunset side being somewhat preferred). In some maps, valleylike areas were discernible by this shading within a particular mountain chain or grouping. Such were among the guides that may have aided Galileo and his engraver.

There can be little doubt that, as a mathematics professor and astronomer, Galileo had been exposed to these conventions many times. The mentioned atlases were among the most widely bought and sold items of the age among the scientific cognoscenti. Ortelius's work, for example, was viewed as one of the great literary-scientific events of the era, going through dozens of editions before 1600, not counting pirated, plagiarized, and other unofficial versions. Maps were among the necessary possessions of any educated man in the early colonial period, especially scientists. Galileo's own link to fields such as navigation and magnetism (he was an avid reader of Gilbert's work), his long-term interest in the tides, and his many years spent at the university of Padua, then an intellectual nexus for scientists of every type, must have brought him in contact with maps many times. Indeed, while a young man in Florence, he had been called to the local academy to speak on the geography of Dante's *Inferno*.[18]

The absence of names in Galileo's portrait and description of the Moon may seem strange: given his many drawings and his sense of being the first lunar "explorer," wouldn't the impulse to name this surface have naturally occurred to him? Didn't he after all give titles to the four Jovian moons he discovered, satellites he baptized the Medicea Sidera (the Medici stars) in honor of the patron whose support his book was intended to procure? He named four moons on which he could see nothing, but left blank and naked the Moon itself, on which he mapped a fabulous plurality of features. The only explanation for this is the one already given. Galileo wanted his images to be perceived—and probably perceived them as such himself—as recorded sights transcribed to paper. This means they were to have a threefold function: demonstrative by acting as proofs of his witness; epistemological, by their power to fulfill the requirements of "evidence" and "discovery"; and didactic, by teaching a new visual language with which to comprehend lunar reality. To achieve all this, Galileo's pictures had to be convincing as observations, seemingly peeled from the eye yet legible by the requirements for visual literacy to the era for which they were produced.[19] Galileo thus provided a map, an artwork, and a visual manual all at once. If the "Medici stars" were the sign of his dependency on the existing system of patronage, to the Moon he gave a new discourse that effectively turned this relationship on its head.

THE GALILEAN LEGACY: AN INTERLUDE OF CONFLICT

In the decade and a half after *Sidereus nuncius* appeared, the Moon became a frequent, if conflict-ridden, subject of discussion and portraiture. This renewed lunar interest revealed the Galilean legacy of bringing together art and science. Although Galileo's work had married Renaissance aesthetic techniques to the production of scientific evidence, it also set loose a series of debates that echoed from one end of the century to the other and that played itself out in astronomical treatises on the one hand and in religious paintings on the other.

This does not mean that Galileo's work was everywhere the subject of challenge or dispute. On the contrary, as the examples of Harriot and Kepler make plain, *Sidereus nuncius* was widely accepted, especially after the telescope became generally available and the relevant discoveries could be confirmed. Moreover, Galileo's images became the source of constant imitation, much of it clumsy and stylized, that lasted for decades. Yet his celebrity was equally due to his pointed rejection of the entire system of traditional Aristotelian cosmology, based on the notion of perfect, homogeneous, crystalline bodies. This system could not accommodate a rough, mountainous, Earth-like Moon. The entire presumption of a single lunar substance divided into "rarer" (reflecting) and "denser" (nonreflecting) parts had to be abandoned, as did that of the purity of the Moon and its religious associations.

This aspect of the larger controversy over *Sidereus nuncius* (which also involved the discovery of the Jovian moons, the shape and nature of Saturn, and other issues) was most keenly rejected by Jesuit scholars in Italy. It was their effort to shore up a spotless lunar conception against the Galilean one that found expression in the paintings of certain prominent Italian and Spanish artists. The story of this debate and its artistic influences has been told in excellent detail by Eileen Reeves and will not be repeated here.[20] A brief summary would show that a number of artworks completed between 1612 and 1619 restaged the conflict between the Scholastic and Galilean Moons by taking up a single subject: the Immaculate Conception, as interpreted from the vision of *Revelation* 12:1–4. In this vision, "a great and wondrous sign appeared in heaven: a woman clothed with the Sun, the Moon beneath her feet, and on her head a crown of twelve stars. She was about to give birth and in the agony of her labor cried out." As discussed earlier in chapter 4, one strand of textual commentary had long identified the woman with the Holy Virgin and the Moon beneath her feet with a pure, homogeneous body aglow with light. During the later medieval period, this image had itself been purified by the substitution of the Immaculate Conception for the "labor pains" in the original, and the notion of a lunar body chaste of all diversity and terrestrial substance received no small support from the interpretations of Averroes and Aristotle that claimed a crystalline, even diaphanous nature for the Moon.

To defend the virginal lunar substance, painters such as Francisco Pacheco, Diego Velásquez, and Francisco de Zurbarán produced images that showed Mary in flowing robes, slender and demure, hands gently folded, and feet posed upon a glowing, translucent white orb without a single spot. These paintings were a direct response not only to *Sidereus nuncius* but also to its artistic equivalent painted by Galileo's longtime friend and former fellow student, Lodovico Cigoli, who had rendered the vision of *Revelation* 8 upon the dome of Santa Maria Maggiore in Rome. Cigoli's image has often been discussed in recent years. It has been noted more than once that the Virgin Mary as portrayed in this work qualifies as the very embodiment of Earthbound substance: massive and heavy set, gazing upward, and standing with sufficient weight upon a cratered, stonelike Moon to flatten it into an elongate boulder. Cigoli seems to have gone out of his way to make the lunar body as solid, pockmarked, and "impure" as possible; his Moon, taken out of the skies and turned into a pedestal, is even more terrestrial than that of Galileo.

What of the images produced by astronomers or printed in books for the scientifically minded? It is curious to find that, in this realm at least, no progress was made in creating pictures of the Moon for more than two decades. The first book to follow Galileo, G. C. La Galla's *De phenomenis in orbe Lunae* (1612), simply republished the images from *Sidereus nuncius.* Both books, in fact, were printed on the same press and probably used the same plates.[21] In the images that subsequently appeared up until the 1630s and 1640s, one finds a stylization of Galileo's pictorial language often rewedded to the mathematical tradition by the addition of designating letters. Versions by P. Christoph Scheiner (drawn in 1614), Charles Malapert (1619), Giusèppe Biancani (1620), and Christopher Borri (1627) actually show a progressive decay or regression from Galileo's cartographic naturalism. They turned the Moon into either a diagram with surface features that looked like spots and simple circles or a baroque decoration without topographical character. Scheiner's drawing, for example, employs a different pattern of stippling for each of the lunar maria, shows no craters, and resembles an ornate wall motif (fig. 8.3, upper image). In Biancani's hands, the lunar surface is returned almost entirely to the mathematical tradition (fig. 8.3, lower image). Borri's image shows the lunar maria as crudely outlined blotches, each labeled with a letter and bordered by a rising line of bubblelike craters. Later images drawn by Francesco Fontana in 1629–1630 added new features, such as the explosion craters Copernicus/Kepler and Tycho, and sketched the maria in more detail. But these, too, are fanciful and decorative, scattering little cell-like craters randomly across the surface.[22]

This failure in post-Galileo astronomical illustration returns us to the realm of art. Cigoli's image of the Moon cannot be called naturalistic; it is more of a materialistic allegory. Far more impressive from a realistic point of view is the lunar orb presented in Adam Elsheimer's *Flight into Egypt,* housed in the Alte Pinakothek museum in Munich. Considered by many to be Elsheimer's masterwork, the painting was apparently begun in 1609 (this date is inscribed on the back) and was finished sometime in 1610, being

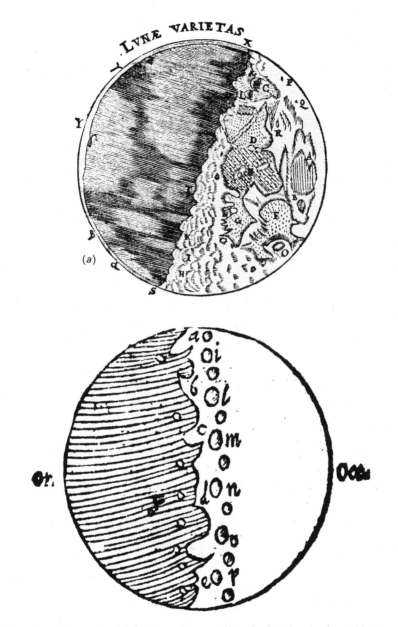

Figure 8.3. Two versions of the Moon drawn within the first decade after publication of Galileo's Sidereus nuncius (1610). The upper figure is an engraving from Christoph Scheiner's Disquisitiones mathematicae (1614). The lower image is a woodcut from Sphaera mundi (1620) by Giuseppe Biancani. Courtesy of Ewen Whitaker.

found in his studio upon his death in December of that year. There is another reason to suppose this date of completion. Elsheimer was living and working in Rome at the time. There is little doubt that he encountered Galileo's *Sidereus nuncius* soon after it appeared, possibly because of the sensation surrounding its lunar imagery.

Flight into Egypt is a magnificent evocation of a nighttime scene deepened by narrative complexity. There are no fewer than five sources of light in the midst of the shrouding darkness. Two of these bear the burden of the story: a small fire, around which huddle the shepherds; and nearby, a torch in Joseph's hand that illuminates the holy family during a moment of intimacy. These punctuations of light are echoed in a brilliant full Moon breaking through the clouds and in the Moon's reflection in a nearby river or lake. Piercing the black heavens where the clouds have parted are dozens of stars and a luminous strip of smaller stars that form the Milky Way. Elsheimer's Moon and its reflection display all the major lunar maria, even more distinctly than do Van Eyck's daytime images. In addition to the Milky Way, which is not in its proper position, the painting shows a number of constellations, most notably the Big Dipper (in the upper right corner). The depiction of the Milky Way as a band of stars and not as a cloud or mist as in most accounts seems to be direct evidence of Elsheimer's familiarity with Galileo's book.[23]

THE NEXT ADVANCE

For the most part, the lunar images produced for "science" in the first two decades after Galileo were inferior to those created by Cigoli and Elsheimer for "art." Following Galileo's lead, Scheiner, Malapert, and Borri placed north at the top in their lunar pictures—even though this convention had not yet been established on terrestrial maps—and generally provided more drawings of half- and quarter-Moon phases than of the full Moon. However, the use of pictorial conventions common to terrestrial maps is far more tentative in these images than in those of Galileo; it is also for this reason that they appear so crude and schematic in comparison. For nearly twenty years, astronomy struggled to catch up with Galileo in its representational powers, to absorb and then surpass the pictorial rhetoric he established. It would require a true artist, guided by scientific eyes, to accomplish this feat (as will be shown in a later chapter).

The Moon rose anew in seventeenth-century image making because of a central difficulty for colonial exploration: the inability to determine longitude with any degree of useful precision. Unlike latitude, which could be estimated relative to the height of a fixed point in space (e.g., the North Star), east-west position was complicated by the Earth's rotation. During much of the sixteenth and early seventeenth centuries, it was thought that changes in the downward angle (declination) of the compass needle might reveal systematic patterns across the Earth's surface, thereby providing a means for estimating east-west position. When this was proven wrong (the Earth's magnetic field varies in complex fashion and also changes over time), most scientists realized that longitude had to be calculated on the basis of time.[24] The problem involved finding a way by which time could be recorded at exactly the same instant yet at different locations. Galileo's observations of the lunar surface suggested a technique, which would involve noting the exact times when certain features—mountains, for example—went into shadow or emerged into sunlight during lunar eclipses. For this, however, a highly accurate map of the Moon's surface was essential. It was for this purpose that the next phase of lunar mapping and naming was undertaken by several different teams simultaneously.

The problem of longitude was very much tied to the political affairs of Europe. It was especially linked to the ongoing competition for trade routes, colonies, and resources; to naval conflicts at sea; and to the internal struggles of various nations to gain full command over the territories and borders within their grasp. This last link should not be underestimated. One tends to think of the colonial era as a time when the new nation-states of Europe were engaged in outward expansion. This process, however, was part of these states' efforts to map, measure, and control the space within and along their borders. In blunt terms, "knowledge of the territory became inseparable from the exercise of sovereignty."[25] Accurate cartography, the precise ordering of space in plotted form, had become coterminous with the waging and winning of wars, the establishment of a centralized administration, and the ability to exploit and mobilize resources of all kinds—the very power, in other words, to create and solidify a nation-state.

The inability to determine longitude was a very large gap in this growing power of spatial command. It is hardly unexpected that a ruler such as Phillippe III of Spain would offer a one-time reward of six thousand ducats plus a life pension of two thousand ducats to anyone who could solve this problem. No doubt this provided some incentive, especially for Catholic scientists. But the historical rewards must have seemed far greater than this—they promised the immortalization of one's name, as both a national and an international hero, during a period when the mixture of rivalry and cooperation among scientists sometimes matched that among nations. The Moon, as a new celestial compass, was drawn into this larger circumstance and through this, finally gained its nomenclature.

9

Return of the Text

LITERARY EXPLORATIONS OF LUNAR GEOGRAPHY

IMAGINING THE OLD IN THE NEW

Galileo's *Sidereus nuncius* enjoyed a tremendous reception throughout Europe and became almost immediately the subject of news reports, poems, plays, satires, and of course, imitations. The disturbance caused by this slim volume, packed as it was with unprecedented tales from the stars, led to a vast cultural mastication. What emerged in the domain of astronomy was a small series of works that did little to advance lunar studies. Elsewhere, however, something very different took place: a host of new works arose, mixing opportunism with literary ingenuity to refound an entire genre of writing. The literary impulse seems to have been a crucial means by which the new universe of Earth-like planets and countless stars was made part of the ambient culture.

The theme of the cosmic voyage in the literature of the seventeenth and eighteenth centuries has been discussed in fascinating and erudite fashion by Marjorie Hope Nicolson in what have become classic studies in the history of ideas.[1] Rather than try to add to this work or improve upon it, I would like to give it a different interpretive cast in light of the greater history of the Moon on Earth. Nicolson perceptively notes that among all discoveries made by astronomers between 1500 and 1700, it was the determination of the true nature of the Moon that proved to be "the most readily comprehensible to the layman,"[2] and thus to writers and poets. But the reasons for this comprehensibility and how it was used were more revealing of the historical moment than has been allowed.

AN EARLY EXAMPLE: *ORLANDO FURIOSO*

Lucian's satires (discussed in chapter 2) were not the first word in the genre of celestial travel. Whether by dream, whirlwind, or flying chariot, the journey to heaven was a device that had been used since the classical Greek period. Early prototypes can be found in Plato—the ascent and descent of human souls through the heavenly spheres in *Timaeus* and the "flying chariot" of the *Phaedo*—and in Cicero's beautiful paean to the Republic in *Somnium Scipionis* (Dream of Scipio). We have also seen that a visit to the Moon took place in Antonius Diogenes' *Wonders beyond Thule,* where links between the lunar body and the fantastic were openly made.

During the Renaissance, when these works became part of the university curriculum or an essential element of every scholar's reading, similar voyages arose within the writings of literary authors. One of the most remarkable and enjoyable examples is found in Ludovico Ariosto's epic romance, *Orlando furioso* (1532), a work of enormous popularity and influence that was translated into every European vernacular before the end of the sixteenth century. The setting for the voyage involves a fated meeting between Duke Astolfo and St. John the Baptist at the top of a distant, wondrous, and paradisial hill. St. John tells the duke he will aid him in his mission to help the great knight Orlando regain his lost senses, which have been exiled to the Moon in a jar as a result of Orlando's madness brought on by excessive love and jealousy. With St. John as guide, the two ascend to the lunar surface in a four-horse chariot, where they find a world ripe with both marvels and dense familiarities:

> 'Twere infinite to tell what wondrous things
> He saw that passed ours not few degrees,
> What towns, what hills, what rivers, and what springs,

What dales, what palaces, what goodly trees;
But to be short, at last his guide him brings
Unto a goodly valley where he sees
A might mass of things strangely confused,
Things that on Earth were lost or were abused.[3]

These "things . . . lost or . . . abused" include what was squandered in life needlessly and intentionally, whether this be "the precious time that fools misspend in play" or "those fond desires that lead us oft astray." There are aspects of Dante here, metaphoric images turned literal in their transposition to another world: "a wondrous hill of bladders" is the site for "pompous crowns and scepters" of kingdoms past; a "great store of baited hooks with gold" are the mimetic material of "gifts that foolish men preferred" (pp. 294–295). These castings from the Earth below, however, are but the beginning of Astolfo's lunar experience. After receiving the vessel containing Orlando's wit, which is safe and intact, he accompanies St. John to an unpleasant river, where he encounters the Three Fates "that the thread of life do spin" and Father Time, "wondrous old," who is carrying heaps of names that he then casts "into this stream, which Lethe we do call" (p. 304). Some of these names, one learns, are rescued by swans and placed in a sacred church upon a hill. Such reprieve, the apostle informs us, occurs for the few because of how "historians learned and poets rare/Preserve them in clear fame and good report" (p. 306). The journey then comes to a close, as St. John bemoans the lack of such writers at present and transports himself and the duke back to Earth.

Ariosto's epic thus weaves together many ancient strands of lunar imagery. Some threads have been plucked from Lucian (the chariot and the earthly landscape), some from Pythagorean thought (the Moon as a purgatorial domain), and some from folklore (the site of madness), astrology, classical myth, and other sources. In all of these cases, the Moon's surface is posed as an alter-Earth defined by its influences on terrestrial destinies. "Know first," says St. John to a wondering Astolfo, "there cannot wag a straw below . . . but that the sign is here" (p. 305). Yet we are also told that "historians learned and poets rare" can rescue fame, thus that the power of literature can counter the force of the Fates, Time, and the river Lethe all at once. Ariosto, no doubt, is making a plea for his own ambition as an author, but he is doing so by rejuvenating the age-old concept of *chthon* and *antichthon*. What the Earth has

lost and wasted, the Moon reclaims. If love is madness, the lunar journey re-
stores sanity by recasting the familiar in the form of ironic wonders, thereby
providing moral insight. For Ariosto, as for Lucian, the Moon constitutes a
geography of satirical reflection: whatever light we receive from it has had its
origin "below."

BEN JONSON'S *NEWES*

Decades before Gilbert, Harriot, and Galileo, the terrestrial view of the Moon
was established as a literary theme. Seventeenth-century accounts of voyages
to the Moon mainly continued in this vein, with various additions or amend-
ments. Such accounts underwent a veritable explosion between 1600 and
1650 and built their discourse by borrowing generously from three sources:
Galileo's *Sidereus nuncius;* fantasy-utopian literature; and explorational accounts of the New World, such as Harriot's own *Briefe and True Report.*

One of the earliest examples and one of the most intriguing from a
historical point of view was a masque[4] by Ben Jonson, written in the winter
of 1619 and performed at least ten times by March of the following year.[5]
Newes from the New World Discover'd in the Moone has been commonly offered a
minor, even trivial place in Jonson's larger corpus in this genre, which includes twenty-eight individual plays.[6] More recently, however, *Newes* has
been skillfully analyzed in terms of political events of the time and how these
achieve a subtle and complex rendering in what otherwise appears to be
bawdy and arcane drollery.[7] The political situation during the six months
prior to June 1620 was an extremely heated one. The outbreak of the Thirty
Years War was imminent because of heightened tensions between Protestant
Bohemia, along with its allies in Holland, and the Catholic Holy Roman
Empire backed by Spain. A complicated play of dominoes ensued and, as the
Dutch understood all too well, teetered on the edge of war, which could be
avoided only if a sign from England was forthcoming to indicate it would
support the Protestant cause against Spain. James I, then king, stood by and
did nothing, proving himself an inept judge of existing realities. From the
fall of 1619 onward, he was continually assaulted by advice from all corners
and was reminded of his "plight" via many carriers of the news, both domes-
tic and foreign. Because of this, James avidly disliked all types of journalism.

His obstinate refusal to act might even be seen as childish resistance to this momentum of public expression.

Jonson's masque opens with a parody of news gatherers and sellers. There are three such personages: a printer ("indeed, I am all for sale, gentlemen"); a chronicler ("And I am . . . to fill up my great book, which must be three ream of paper"); and a factor, i.e., one who mails letters containing news to a select clientele ("I have my Puritan news, my Protestant news and my Pontifical news").[8] These three are accosted by two heralds, who bring news from a more distant place. To this announcement, the printer makes an interesting comment: "Oh, by a trunk! I know it, a thing no bigger than a flute case. A neighbor of mine, a spectacle maker, has drawn the Moon through it at the bore of a whistle and made it as great as a drumhead." To which the chronicler retorts, "Tut, that's no news; your perplexive glasses are common" (p. 295). Placed in the mouths of clowns, such words reveal how the Moon and efforts to draw it are no longer front-page material. Galileo's work has become fully integrated into public consciousness throughout Europe within less than a decade. Despite such statements the heralds persist in their announcement:

> 2*nd Herald.* Certain and sure news—
> 1*st Herald.* Of a new world—
> 2*nd Herald.* And new creatures in that world—
> 1*st Herald.* In the orb of the Moon—
> 2*nd Herald.* Which is now found to be an Earth inhabited!
> 1*st Herald.* With navigable seas and rivers!
> 2*nd Herald.* Variety of nations, polities, laws!
> 1*st Herald.* With havens in't, castles and port towns!
> 2*nd Herald.* Inland cities, boroughs, hamlets, fairs and markets!
> 1*st Herald.* Hundreds, and wapentakes [subdivisions of English counties]! Forests, parks, cony-ground, meadow-pasture, what not?
> 2*nd Herald.* But differing from ours. (pp. 296–297)

The irony of the last line surpasses even Lucian. The factor, upon hearing the "news," immediately wants to know "What inns or alehouses are there?"

Jonson, however, is warning his English audience to side with their king. What does James believe? That there is nothing really new to be learned from abroad. The theme of the futility of "false news" and its supposed desire to befoul or mislead the court runs throughout the masque in a host of subtle ways.[9] In a later portion of the work corresponding to the antimasque (in which mocking effects are abandoned for more direct flattery of the king), Jonson introduces a group of "Volatees," who are announced to James as "a more noble discovery worthy of your ear . . . a race of your own . . . , who, rapt above the Moon far in speculation of your virtues, have remained there entranced certain hours with wonder of the piety, wisdom, majesty reflected by you on them from the divine light."[10] The Volatees enter, shaking off the icicles that they "contracted in coming through the colder region," to dance and sing the praises of the king, to "say but James is he" who stands with the Sun as source of light and truth.

The "false news" from the continent, full of the "lunacy" that would beg military intervention, is replaced by the real tidings of the moment—that the king is a shining beacon of "pure harmony" and that all who oppose his views place themselves on the side of the "colder region" and therefore in immediate danger of James's wrath. Though a surrogate in part, Jonson's Moon also carries the concrete reality of threat: it is to this distant and frozen domain that all bearers of improper tidings will be exiled. This is the ancient Peripatetic idea of the Moon's "cold nature" translated into literary utility. In Jonson's indiscriminate usage, Galileo and Aristotle hold hands.

KEPLER'S *DREAM*: AN INSOMNIAC VISION

Named after Cicero's *Somnium Scipionis,* in which the heavens are spread out as a moral-theological tapestry, Kepler's *Somnium* (The Dream) appeared in 1634, four years after its author's death. It was a dream work itself, one that only death could wake into public existence. As a form of literature, it suffers from this fitful gestation, being a quaint and not wholly successful attempt to focus an astronomical eye of musing intent on the Moon. The general context is anything but scientific, as the first few lines make evident:

> In the year 1608 there was a heated quarrel between the Emperor Rudolph and his brother, the Archduke Matthias. Their actions universally recalled precedents found in Bohemian history. Stimulated by the widespread public interest, I turned my attention to

reading about Bohemia, and came upon the story of the heroine
Libussa, renowned for her skill in magic. It happened one night that
after watching the stars and the Moon, I went to bed and fell into a
very deep sleep. In my sleep I seemed to be reading a book brought
from the fair.[11]

These lines are spoken by Duracotus, narrator of the tale, who hails from
Iceland, "which the ancients called Thule." Kepler's Moon is thus very much
a Renaissance orb, combining elements of the present (politics), the medieval
(history), and the ancient (literature and geography).

Duracotus informs the reader that he was sold by his mother to a passing
fisherman and wound up, while still a youth, at Tycho Brahe's observatory on
the island of Hven. He was then transported upward by a daemon to "the
island of Levania" (derived from the Hebrew word for "white"), whose road is
arduous and seldom open. It is this daemon—a narrator within a narrator—
who delivers the main portion of *Somnium,* which turns out to be an extended
discussion, often in astronomical and mathematical terms, about what a
trained observer is likely to see on the lunar surface. Many phenomena are described: the lunar natives, Privolvans and Subvolvans, who inhabit the light
and dark hemispheres, respectively; hot and cold regions and how they alter-
nate; the Earth (called "Volva"), its motions, and its cycle of days, months,
and years; and the appearance of the Sun, planets, zodiac, and other stars.

It is difficult to tell when Kepler is being wholly serious or merely
fanciful. "Relief from the heat in the Subvolvan hemisphere is provided
chiefly by the constant cloud cover and rain," he states in one place, "which
sometimes prevail over half the region or more" (p. 28). Kepler's footnote to
this sentence tells us, "The same statement is made by José de Acosta about
regions in the New World" (p. 135). Fanciful or not, Kepler's Moon is more
fully another Earth than any version to date. This becomes even clearer in one
of the most remarkable paragraphs in the entire work:

So far as its upper, northern part is concerned, Volva [the Earth] in
general seems to have two halves (153). One of them is darker and
covered with almost continuous spots . . . the shape of [which] is
hard to describe. Yet on the eastern side it looks like the front of the
human head cut off at the shoulders (158) and leaning forward to
kiss a young girl (159) in a long dress (160), who stretches her hand

back (161) to attract a leaping cat (162). . . . In the other half of Volva the brightness is more widely diffused than the spot (166). You might call it the outline of a bell (167) hanging from a rope (168) and swinging westward. (p. 24)

These descriptions and the footnotes attached to them strongly suggest that Kepler wrote this while gazing down at a map of the Earth, imagining likenesses in the same manner that mankind had long done while looking up at the lunar face. The "two halves," Kepler says in his note 153, refer to the Old World and the New. The head is Africa (158), the young girl Europe (159); her dress (160) flows out from Thrace and the Black Sea region, while her hand, Britain (161), reaches for the feline shape of Scandinavia (162; p. 114). The bright area of Volva's other half is North America (166); the bell is South America (167) dangling from the rope of Central America (168).

In this brief paragraph, Kepler has indeed put himself upon the Moon to look back (and down) at the Earth with imagination, knowledge, and humor. He has done this, moreover, by using geography as a basis for poetic imagery in a way that recalls, and may well satirize, the tortured and overly complex images that others, such as Albert the Great, had constructed out of the lunar maculae (a lion with a tree on its back, against which leans a man, etc.; see chapter 5). Note, however, that there is a distinct hierarchy in the imagery that Kepler has chosen. Europe is given much detail as the young "beauty" of continents; South America is next, though reduced to the massiveness of a single, suspended figure. Yet North America, newest portion of the New World, remains entirely amorphous: it "cannot be likened to anything" (p. 24) because it remains mostly unexplored and thus without a true form on the world stage. If Kepler's *Somnium* failed as a "complete geography of the Moon," it was nonetheless a success in terms of how it embodied contemporary sensibilities about the Earth.

Somnium contained a final section called a "Geographical or, if you prefer, Selenographical Appendix." This appendix is written as a letter and contains thirty-four axioms of lunar geography, plus (should we be surprised?) a Latin translation of Plutarch's *On the Face Which Appears in the Orb of the Moon*, which Kepler had apparently studied while writing portions of *Somnium*. Like the more serious, technical parts of the dream itself, this appendix presents a central element of Kepler's imagined truth about the Moon. Many of the axioms, he states, have resulted from his own telescopic observations,[12] which have revealed to him the following details:

> If you direct your mind to the towns on the Moon, I shall prove to you that I see them. Those lunar hollows, first noticed by Galileo, chiefly mark the Moonspots . . . depressed places in the flat area of the surface, as the seas are among us. But from the shape of the hollows I conclude that those places are . . . swampy. And in them the Moon-dwellers usually measure out the areas of their towns for the purpose of protecting themselves from the mossy wetness as well as from the heat of the Sun, and perhaps even from enemies. The design of the fortification is as follows. They drive a stake down in the center of the space to be fortified. To this stake they tie ropes which are either long or short, depending on the size of the future town. The longest I have detected is five German miles. With this rope fastened in this way, they move out to the future rampart's circumference. . . . Then the entire population assembles to do the digging. . . . They take all the excavated material inside some towns. In others, they have built partly outside and partly inside. . . . Thus the result is that not only is the ditch pushed down quite deep, but also the center of the town looks sunken like a chasm, as though it were the navel of a puffed up belly. . . . When it overflows with water, [the chasm] becomes navigable. When it has dried out, it can be crossed as a land route. (p. 151)

For Kepler, no less than for other astronomers, the Moon was something to be comprehended through the filter of "territory." The implications of Gilbert's map and the concreteness of Galileo's images met in these pages with a specificity that can hardly be called fictional—that is, if one thinks of everything the seventeenth century had come to associate with the idea of the New World.

GODWIN'S *MAN IN THE MOONE* AND CYRANO'S *L'AUTRE MONDE*

Another early work wove similar tales of lunar and earthly voyages but in a more skillful, unified, and humorous way that made it a publishing success throughout Europe. This work drew less on Cicero and far less on technical astronomy than it did on Harriot's *Briefe and True Report,* both in its brevity (a few dozen pages) and in certain ingredients of its narrative. Francis Godwin's *The Man in the Moone: Or a Discourse of a Voyage Thither by Domingo Gonsales, the Speedy Messenger* (1638)[13] plainly reveals Galileo, utopian literature, and colonial accounts in its title. Like Kepler's *Somnium,* this was a posthumous work (Godwin died five years earlier), probably written in the late 1620s or early 1630s. It used Francis Bacon's *Sylva sylvarum* (1626) as a reference, as is evident in the borrowed "flying chariot" drawn by a flock of enthusiastic, migrating geese, which carries Godwin's apocryphal Spanish "hero" to the Moon. The book is a partly humorous, satirical account of an undeserving opportunist (tiny in stature, cowardly by nature) who, after being tossed from one life circumstance to another, across various and sundry parts of the globe, lands on the island of St. Helena. From there he is transported to the lunar surface, where he meets various inhabitants and their rulers, who convert him to a better outlook, after which he returns to Earth, landing in China. The story ends with the narrator hoping he might one day return to Spain so that "by inriching my Country with the knowledge of hidden mysteries, I may once reape the glory of my fortunate misfortunes" (p. 48). With such hope for colonial fame, Godwin's Gonsales, like Kepler's daemon, reports that the Moon has its seas, lands, islands, and other features and describes a catalog of items and experiences that serve no less as a "map" of his discovery than did Harriot's report. In the opening preface, moreover, Gonsales advises us:

> In substance thou hast here a new discovery of a new world, which perchance may finde little better entertainment in thy opinion, than that of Columbus at first . . . Yet [whose survey] of America [hath since been] betray'd unto knowledge soe much as hath since encreast into a vaste plantation . . . That there should be Antipodes [people living on opposite sides of the Earth] was once thought as great a Paradox as now that the Moon should bee habitable. But the knowledge of this may seeme more properly reserv'd for this our discovering age: In which our Galilaeusses, can by advantage of their spectacles gaze the Sunne into spots, & descry mountains in the Moon. (p. 2)

This is not the stuff of satire or fancy; Godwin's narrator shows us that he is fully in touch with the events of the day. However fictional his succeeding tale may be, these words make him akin to a historical keyhole, our sign that on the plains of discovery, Columbus and Galileo were equals. Godwin's tale gives us the Moon as a territory to be visited, traded with, learned from, and possibly exploited, just like the "far side" of the Earth.

Gonsales strives to emulate Harriot's *Briefe and True Report.* He describes such things as food, metals, precious stones, language, ethnography. He also notes that the Moon's inhabitants are able to discern at birth who among them will be wicked or imperfect and that these undesirables are then "vented" onto the Earth through "a certaine high hill in the North of America, whose people I can easily beleeve to be wholly descended of them, partly in regard of their colour" (p. 40). The inversions here are many and not to be missed. The mouth of England's enemy heralds the "discovery" that the natives of England's newly claimed New World are, in fact, dangerous rejects who have been sent down to Earth. By way of the Moon, North America is proposed as a colonial prison.

Gonsales—actually Godwin—performs another trick of reversal, the same as that achieved by Kepler in *Somnium:*

> [W]hereas the Earth according to her naturall motion (for that such a motion she hath, I am now constrained to joyne in opinion with Copernicus) turneth round upon her own Axe . . . I should at the first see in the middle of the body . . . a spot like unto a Peare that had a morsell bitten out upon the one side of him; after certaine [hours], I should see that spot slide away to the East side. This no doubt was the maine of Affrike. Then should I perceive a great shining brightnesse to occupy that roome, during the like time (which was undoubtedly none other than the great Atlantick Ocean). After that succeeded a spot almost of an Ovall form, even just such as we see America to have in our Mapps. Then another vast cleernesse representing the West Ocean; and lastly a medly of spots, like the Countries of the East Indies. (p. 22)

It seems unlikely that Godwin knew of *Somnium.* Certainly, the images he chose to employ for the terrestrial "spots" were different from those of Kepler. What both works imply, therefore, is a readiness to see the Earth itself in a new light—this is what the vision of the Moon, as a planet with a true geographical surface, offered.

Godwin's work proved to be immensely popular in its day, both in England and on the continent, and was continually reissued well into the eighteenth century in various translations.[14] It was even once included as a type of travel literature in a Hakluyt-type work called *View of the English Acquisitions ... in the East Indies,* assembled by publisher Nathaniel Crouch in 1686, possibly on the impression that it was a story of an actual voyage. Godwin's fiction, moreover, was undoubtedly one of the chief inspirations for Cyrano de Bergerac's own *L'autre monde: Les etats et empires de la lune; les etats et empires du soleil* (1649), a much longer work with space to spare for lush satires on everything from the Book of Genesis and Pliny to "modern" education and French scientific society. Cyrano, in fact, may be the most brilliant and complex of all seventeenth-century writers in this proliferating genre of lunar *voyageurs* and inspired some famous latter-day authors of such tales, especially Jules Verne.

L'autre monde opens as a dialogue, reminiscent of Plutarch, between ac-quaintances who are discussing whether the Moon might be an inhabited world like the Earth. Cyrano's hero, the narrator, is ridiculed for stating as much and after returning home, determines to prove his point by an actual journey to the lunar surface. His voyage begins when a group of soldiers accidentally sets off fireworks placed in a chariot that the narrator is prepar-ing for flight, thus producing the first rocket ship. Reaching the Moon, the narrator discovers that he has landed in Paradise, where great rivers, stones, scented flowers, and perpetual spring all speak their reality to him in soft, mellifluous tones. The traveler feels the years roll off him; he is returned to the vigor and beauty of a fourteen-year-old. The ancient notion of a lunar purification is enacted; Cyrano is well informed on classical travel literature as well as cosmology.

The narrator next meets his guide, Elias, who plays a role like that of St. John to Astolfo in *Orlando furioso.* Elias explains many things about the Moon, including the fact that only six people have ever been allowed by God to venture here (Adam, Eve, Enoch, St. John, Elias himself, and now Cyrano's hero). He also discusses the Tree of Life and the Tree of Science, which are planted side by side. As for the latter, "Its fruit is covered with a bark that makes an ignoramus of anyone who tastes it, yet beneath whose coating are the spiritual virtues of the learned, there for the nibbling. . . . At some point after chasing Adam from this happy world, God himself, fearing that Adam might well find his way back, rubbed his gums with this bark."[15] The narrator himself, stricken by hunger, forgets to shell the fruit and bites into it, fortunately deeply enough to attain a few drops of the salubrious juice.

Many adventures succeed these events and are interspersed with discussions between the narrator and his guide (who takes different forms) about the nature of the lunar inhabitants, their history and habits, their language, and the views they hold of their Moon, the Earth. These inhabitants are huge, use all four appendages for locomotion, and speak through musical sounds and varied trembling of their limbs. There are numerous ironic and satirical inversions: Cyrano is happily updating Lucian in his parody of the "wonder tales" then so prevalent in the literature of New World exploration (Harriot being a notable exception). Often there is bitter insight hovering in his diagnoses, as when he caricatures explorers' accounts of "discovering" a new people in distant lands:

> They said therefore . . . that without any doubt I was the female species of the queen's small pet. Thus, I displayed this or that quality or aspect. . . . A certain burger who kept rare beasts requested of the magistrates that I be handed over to his charge until time when the queen summoned me to live with my male. As there was no objection, the ruffian then led me away to his lodgings, where he taught me to act the buffoon, to turn somersaults, make faces, and in the evening hours after dinner, he set up a booth and charged admission of all who wished to come see me.[16]

The world for our narrator is thus turned entirely upside down; this perspective continues throughout the latter part of the work. Convicted prisoners on the Moon are "condemned" to a natural death and to burial: the fate of one day being placed in the ground and devoured by worms (instead of being cremated) constitutes a horrible punishment almost beyond imagination. Likewise, lunar males wearing pendants graced with images of the virile member are considered noble gentlemen, in marked contrast to Earth-bound men who parade around wearing swords. "Unfortunate country!" says a lunar speaker, "where the signs of generation are considered ignominious and where those of annihilation are honorable, as if there were something more glorious than the giving of life and nothing more unspeakable than taking it away."[17] The effect is thus to reveal the abominable provinciality, irrationality, and frequent barbarism of so-called civilized society. After all of his adventures and discussions, the narrator returns to Earth completely unchanged, entirely unpersuaded by the arguments he has heard. In the fashion of the day and after the model of antiquity, he labels everything that has happened to him as "marvels." This is where the tale ends. The vast failure of civilization is that it cannot learn even from experience: the New World is there to confirm the old, not to change or expand it.

More than with Godwin or other writers in this genre, the fantastic voyage in Cyrano's hands becomes a vehicle for political, moral, and philosophical commentary. If the outer limit of travel literature is satire of the familiar, *L'autre monde* takes this one step further: the Earth and all its inhabitants, says this author, constitute the true site of "lunacy." Cyrano thus represents one of the more sophisticated literary blendings of ancient and modern imagery associated with the Moon. His work, an updated form of the classical dialogue (over 75 percent of it takes place as exchanges of conversation), brings together the "wonders beyond Thule" tradition of travel writing with its satirical offshoots, the various conceptions surrounding the Earth-like nature of the Moon and the fabled qualities of its inhabitants; much philosophy; and even biblical criticism. Cyrano was an avid follower and friend of the mathematician, astronomer, and philosopher, Pierre Gassendi, who also had decided that the Moon was inhabited, largely on the basis of Galileo's findings, and whose ideas on the failures of human society are abundant throughout this work. *L'autre monde* might itself be read as an ironic title. There is, after all, so much of the Earth in its mapping of social topography.

CONCLUSION

The same year that Godwin's *Man in the Moone* appeared, another more serious effort at popularization was published. This was John Wilkins's *The Discovery of a World in the Moone* (1638). Wilkins, who later became one of the founding members of the Royal Society of London, which embraced "the new experimental philosophy," spreads the Galilean-Copernican gospel in this work. His treatment is earnest, scholarly, and pedagogic, yet it overlaps entirely with the genre of the fantastic lunar voyage.[18] It argues not only that the Moon includes land and sea, "high mountaines, deepe vallies, and spacious plaines," but also "that there is an atmosphaera, or an orbe of grosse vaporous aire" and that, if there are inhabitants, "as their world is our Moone, so our world is their Moone."[19] Wilkins makes a concerted effort to present views of lunar reality as they have been expressed through the ages. His catalogue is selective, aimed at finding support for his main thesis—the habitability of the Moon. He surveys classical authors, medieval commentators, modern writers, and even the Gospels. Finally Wilkins settles on Kepler's own claim for lunar territory in *Dissertatio* (examined previously), and in this choice, shows himself no less the patriot of political astronomy:

> It is the opinion of Keplar, that as soon as the art of flying is found out, some of their nation will make one of the first colonies that shall transplant into that other world. I suppose his appropriating this preheminence to his own countrymen, may arise from an over-partial affection to them. But yet . . . whenever that art is invented, or any other, whereby a man may be conveyed some twenty miles high . . . then it is not altogether improbable that some other may be successful in this attempt.[20]

No clearer statement of the intentions behind creating a lunar geography was ever made; unless it is Samuel Butler's elegant satire on the Royal Society, "Elephant in the Moon" (1660s?), which begins as follows:

A Learn'd Society of late,
The glory of a foreign State,
Agreed upon a Summer's Night,
To search the Moon by her own Light;
To take an Invent'ry of all
Her Real Estate, and personall;
And make an accurate Survey
Of all her Lands, and how they lay,
As true as that of Ireland, where
The sly Surveyors stole a Shire;
T'observe her Country, how 'twas planted;

With what sh' abounded most, or wanted;
And make the proper'st Observations,
For settling of new Plantations.[21]

To say that colonial ambition defined most of the interest shown in the Moon at this time would be absurd—there is more truth to the inverse. Such interest in a "new world" drew upon a host of "old world" ambitions, needs, desires, visions, anathemas, and traditions. From Ben Jonson to de Bergerac and beyond, the uses ascribed to the lunar surface by the seventeenth-century literary imagination were no less of the moment than those of astronomers attempting to fathom the lunar substance and to map and eventually name its features. Later in the century, these uses would even include censorship, cloaked as Protestant modesty in the poetry of Milton's *Paradise Lost* (Book VIII, lines 140–170):

What if . . . land be there [upon the Moon]
Fields and inhabitants: her spots thou seest
As clouds, and clouds may rain, and rain produce
Fruits in her softened soil, for some to eat . . .
    Solicit not thy thoughts with matters hid,
Leave them to God above, him serve and fear;
Of other creatures, as him pleases best,
Wherever placed, let him dispose . . .
heav'n is for thee too high
To know what passes there; be lowly wise.[22]

Such admonitions had little effect upon a century so full of discovery both on Earth and in heaven. Yet they provide one more instance of how the Moon, as a new "marvel" of exploration and hypothesis, served as a historical sponge for sensibilities then percolating through European society. Astronomers, mathematicians, writers, and poets all show that the Moon had undergone a revolution in public consciousness within mere decades after Galileo. From a philosophical object, ripe with essence, the Moon had been transformed, once and for all, into a solid alter ego of the Earth.

10

Efforts from France and Belgium

PEIRESC-GASSENDI AND VAN LANGREN

o the Age of Discovery, the Moon was as much a schoolroom without walls as a landing site for colonial satire. Galileo's images were also not void of pedagogy, even propaganda (as broadly defined). Through his specific brand of naturalism, Galileo and his followers offered a Moon interpreted. The same, however, can not be said for another project that produced portraits of the lunar surface unsurpassed in their realism. Indeed, the very success of this project, which produced images startling even today, may have been the reason for its failure.

THE PEIRESC-GASSENDI PROJECT: NATURALISM AT ITS TRAGIC PEAK

The earliest of several new attempts to map the lunar surface was undertaken by the team of Pierre Gassendi (1592–1655) and Nicolas-Claude Fabri de Peiresc (1580–1637). Gassendi was one of the most eminent French mathematicians and astronomers of the seventeenth century, an avid opponent of both Aristotle and Descartes, a friend of Cyrano de Bergerac, and a powerful influence on the thought and literature of the time. His lifelong friend, Peiresc, on the other hand, was more of an accomplished dilettante or *homme universel,* with sufficient wealth to support avid interests in astronomy, archaeology, natural history, geography, and other areas. Peiresc had foreseen, at the same time as Galileo, the possible importance of using celestial objects for estimating longitude. In January 1628, according to a carefully designed plan, he and Gassendi made meticulous, painstaking observations of a lunar eclipse in Aix-en-Provence in southern France while other astronomers did likewise in Paris (one of whom was Marin Mersenne). This orchestrated effort permitted a precise determination of the difference in longitude between these two sites.

The result encouraged Peiresc to organize a much wider network of observers in Rome, Tunisia, Egypt, Syria, and elsewhere (areas where weather would not be a problem) and to place in their hands a detailed lunar map that would make all observations far more accurate and useful. Success here would set new standards, effectively advancing geography into the realm of geodesy.[1] Also a consideration was the hope of establishing a world reference center in France. Peiresc, long-term advisor to the local Aix government, preferred his city for this site. Political realities, however, would have argued for Paris. In any case, the project could not proceed without a highly accurate lunar map, and the two men thus attempted to construct one.

The project was underwritten to a high degree by the enthusiasm and funding of Peiresc himself. Gassendi, with the encouragement of Galileo (who even supplied telescopic lenses for the purpose), was forced to take over much of the work because of his friend's increasing illness and disability after 1634. Yet Peiresc, even in the years of his decline, held fast to a particular insight: to surpass Galileo, a trained artist had to be commissioned to produce the basic lunar images. Fredeau (noted in Gassendi's diary) was the first, having been engaged earlier to do a portrait of a rare gazelle that Peiresc had shipped from Africa, destined as a gift to the powerful Cardinal Barberini in Rome. Being on site, Fredeau was asked to do a chalk drawing of the full Moon on July 10, 1634. The result, however, was less than satisfying. Sketching an animal present before the brush was one thing; trying to draw accurately, in magnified form, an image whose physical meaning and aesthetic qualities remained unclear was something else entirely. Peiresc recognized that Fredeau was not the man for the job. He then enlisted the services of a more renowned artist, Claude Sauvé (or Salvat). On August 26 of the same year, a second image was produced, also unimpressive. It appears that the weather was foggy, but more that Sauvé, equally unaccustomed to looking through a telescope, was similarly unable to draw what he saw. During the next several months weather remained a problem, but a number of images were created. Then in March 1636, a lunar eclipse offered a special opportunity to detail the advance of the shadow over the Moon's surface. This time the outcome seems to have been better, but for some unrecorded reason, no map resulted. Peiresc and Gassendi sent Galileo several of the images for his opinion. Although the Tuscan found some to be of "reasonable" quality, he also noted that they left out "those extremely long ranges of steep mountains and other clusters of jagged shoals."[2]

Sauvé apparently was dismissed. When Gassendi's diary again mentions the project, it is a full year later. Peiresc has persuaded a still more famous artist, Claude Mellan—said to be "one of the great painters of the century and the most precise engraver in copper yet born"—to produce images of the Moon whose worth would be "remembered for all time."[3] Changeable weather once again imposed an intermittent work schedule. Mellan labored as best he could throughout the fall of 1636, apparently using Sauvé's versions as a starting point, before finishing enough drawings and paintings to yield three engraved images: two quarter phases and one full-Moon phase. Gassendi was duly impressed with the result. He wrote to Galileo (who also knew Mellan) in December, promising a copy of the final result.

One can well see from figure 10.1 the cause of Gassendi's enthusiasm. Mellan's images, with their near-photographic precision, are such a startling leap beyond everything that had gone before—even Galileo's pictures—that we are left almost breathless. Mellan's work is the first since Leonardo that was drawn by a true artist (and engraver) of the first rank. Moreover, Mellan had Galileo as a beginning standard and Gassendi and Peiresc's years of observations and experience with other artists to draw upon. Looking at these images, one feels that Van Eyck would have also admired them. A new stage had been reached: Van Eyck showed the Moon in the hands of art; Leonardo depicted art and observation unmixed; Galileo offered art in the direct (even propagandistic) service of science; and Gassendi, Peiresc, and

Figure 10.1. Engraving of the lunar surface by Claude Mellan, prepared in 1636 but never published. Courtesy of Ewen Whitaker.

Mellan presented art as a domain of expertise *within* science. Taking their cue from geographers of the time, astronomers realized that the making of celestial maps demanded the combined efforts of different areas of knowledge and skill. No less than maps of the New World, lunar maps had to be documents on which were inscribed the powers of high civilization.

Yet the magnificence of Mellan's work passed away almost as soon as it appeared. In June 1637, Peiresc died, and with him went much of the life of the project. Certainly the major funding disappeared, and thus, the ability to produce more than the three engravings already completed. Some years later, hoping that the work would not be wasted, Gassendi sent copies of two of these images to colleagues, notably Johannes Hevelius in April 1644, who was then in the midst of his own lunar mapping program. News of this effort led another interested party, Michael Florent Van Langren in Belgium, to rush his own map into print, thereby becoming the first to offer a complete nomenclature for the lunar surface. Such activity reveals both the widely recognized value of a lunar map and the consequent competition to produce one. In such a climate, it was perhaps inevitable that, without any further work or promotion, the Gassendi-Peiresc-Mellan project would disappear into the shadows of "influence." One of the great failings of Mellan's images was also their success, namely their purely visual character. Unlike Galileo's engravings, which were also naked of writing, Mellan's images did not appear embedded in a textual work; they were unexplained, lacking the legitimizing platform of technical narration. Magnificent yet orphaned, they were closer to art than to science, which remained textually based. As such, they qualified only as raw material. Only a map complete with named features would suffice for determining longitude.

This does not mean, however, that a naming scheme was ignored. Gassendi's correspondence contains the outline of such a scheme, which was never made official but obviously viewed as an important part of the original project. Gassendi's designations are in Latin and remain obedient to the concept of geographical "seas" and "lands," being otherwise of varied character. Like Gilbert, he named a Northern Sea (Boreum Mare), and like Harriot, he recognized a similarity of form between the Mare Crisium and the Caspian Sea. The Mare Humorum, which lies on the opposite side of the visible hemisphere and shares a similar shape, he called Anticaspia after

Ptolemaic convention. The Mare Vaporum he baptized Hecates Penetrale, using the Greek name. On his list of titles, he included Vallis Umbrosa, Rupes Nivea, Amara Mons, and Lacuna, thus calling directly upon terrestrial features (*rupes* means "scarp"; *mons* means "mountain" or "hill"). The combined area of the Mare Serenitatis, Tranquillitatis, Fecunditatis, and Nectaris he lumped together as Homuncio, a region on the east side of Mellan's drawing that crudely resembles a human form slightly tilted backward (no doubt corresponding to the "man in the moone"). The explosive crater Tycho, so prominently displayed on Mellan's image in the lower center-left, Gassendi called Umbilicus Lunaris; the other such crater, Copernicus, located in the upper left, he titled Carthusia, after either the monastery or the area of its location immediately north of Grenoble. Gassendi's system of nomenclature, in its imaginative variability, was therefore intended to project onto the Moon a host of titles derived from classical myth, contemporary geography, and even human anatomy. It was to be (as it largely is now) a Latinate surface, no less than the text of Virgil, the early seventeenth-century maps of North America, or a medical treatise of the time. All of these were "geographic" too, concerned with locations, places, and the contours of visibility.

There is another dimension to Gassendi's system. In his letters to Lobkowitz, Gassendi states with a degree of resignation that he will happily leave to others a more detailed naming of the surface. Lobkowitz takes the cue and writes back with the proposal that "promontories, islands, and valleys" be given the names of contemporary savants. "All our friends will be there," he says with sympathy, "yourself, Peiresc, Mersenne, and Naudé."[4] To Gassendi's system, therefore, was added a canonizing dimension, not merely aimed at the "greats" but also at a circle of friends, whose geography of communal striving, achievement, and disappointment would find its final and eternal resting place in the rugged topography of another world.

History was to prove itself kind to this scheme. Whereas Van Langren would substitute his own memorial canon and Hevelius would abandon this approach altogether, the final arbiter of lunar names, Giambattista Riccioli (1598–1671), would reinstate Lobkowitz's and Gassendi's idea. In 1651, Riccioli gave Gassendi (now in his declining years) and his departed friend Peiresc back to the orb that had absorbed so much of their lives.

VAN LANGREN: THE FIRST TEXTUAL MAP
AND A CATHOLIC MOON

The version of the lunar surface produced by Michael Florent Van Langren (1607?–1675) has been the subject of several studies.[5] Despite the work of Galileo and Gilbert, it is generally regarded as "the first true map of the Moon; that is, it depicted not only the surface shadings as seen at full Moon, but also a large number of topographical features (craters, peaks, mountain ranges) which only become visible at other phases."[6] It should be noted, however, that Van Langren did not seek to produce an "observation," an actual visual portrait of the lunar surface (fig. 10.2). His images were specifically intended to solve the problem of determining longitude. They are true maps, stylized plots of features across a flat plane, unburdened by the demands of naturalism. They are cartographic abstractions whose "science" exists in the accuracy of placement and size of individual features, not in their depiction.

By today's conventions—but even more, by those of mapmaking at the time—the above statement is largely correct: Van Langren's image is the first true modern map of the Moon. (Gilbert's map was more of a reconnaissance version.) This is true for another reason. The lunar surface of figure 10.2 is cartographic because of its dense textuality, because it is surrounded, framed, nested, and invaded by writing in exactly the same way as were mapped regions on Earth. This is a different textuality than is seen in medieval astronomical drawings and diagrams. The medieval illustration was a secondary bearer of knowledge; very rarely could it stand alone without any written context. It was through an art that extended beyond writing (e.g., the works of Van Eyck and Leonardo) that the physical world, including the Moon, obtained a true freedom from words. The telescope helped advance this freedom to an unprecedented degree, making the pure visibility of lunar features the promise or embodiment of technical knowledge. With Van Langren, however, we enter a new stage, where illustration becomes capable of being self-contained because of an equality between text and image. On Van Langren's map, words serve many functions. Names are everywhere, especially those of then-living Catholic royalty. In the corners are quotations from Plutarch, Theodorus, Cicero, Pliny, and Seneca. Below the map is a long confessional narrative of the author's struggles to conceive and finish the

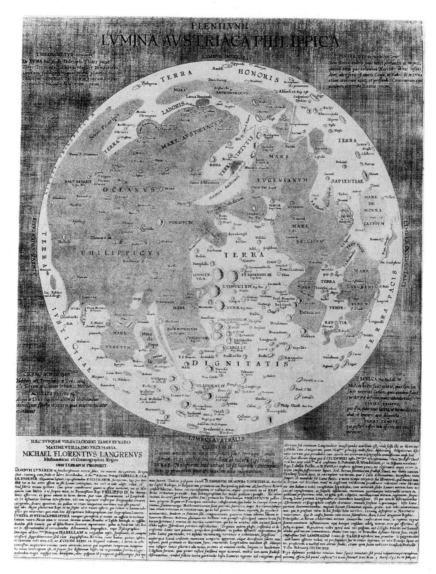

Figure 10.2. *Lunar map published by Michael Florent Van Langren in* 1645. *This map is probably the most complete of several versions prepared by Van Langren. Courtesy of Ewen Whitaker.*

project. A title is also inscribed on the map: *Lumina Austriaca Philippica* (By the Light of Philip of Austria). Labels and creators of place, scholarly references, an extended signature, and obedience to the laws of patronage: all of these roles are played by words within the borders of this map. It is a new Moon that we encounter here, ripe with the touches of seventeenth-century intellectual and political reality.

Little is actually known about Van Langren's life. He came from a family of royal cosmographers and mathematicians, and to these skills he added his own of engraving, engineering, and cartography. Such abilities made him ideally suited to produce the first Earth-type map of the lunar surface. Personally, religiously, professionally, and politically, he was deeply committed to the fortunes of a Catholic Europe led by Spain. His family had moved south out of what is now Holland, a center of long-term Protestant revolt, to the Spanish Netherlands, settling in Brussels. The period 1618–1648 saw the rages of the Thirty Years War, a culmination of religious and dynastic conflicts generated during the late sixteenth century and the period of Spain's downfall as the dominant European power. Spain entered this war in 1621, on the death of Philip III and the ascent of his son Philip IV. By this time, Van Langren—possibly responding directly to Philip III's offer of monetary reward but more likely to the probability of royal favor and ensuing fame—was already working on a method to determine longitude at sea and, like Peiresc, focused his efforts on the Moon.

He presented his idea as early as 1625 to Isabelle, Philip III's sister and the Princess of the Spanish Netherlands. By this time and during the years immediately following, he was at work on several other projects of immediate political and military utility: plans for a fortified port at Mardyck; a map and similar plan for the harbor at Oostende; a map of a canal system linking the Rhine with the Meuse River in Spanish-controlled territory; other maps of Luxembourg, Mechlin, and the duchy of Brabant; designs for a three-shot cannon; and various astronomical and maritime tables. Impressed with such diligence and evidence of loyalty, Isabelle became Van Langren's patron and agreed to write to the king (now her nephew, Philip IV) to request an audience for the young cosmographer. In the intervening months, Van Langren transferred a great deal of his affection to the archduchess, to whom he taught astronomy and whose favor and attentions defined an important part of his life. Eventually granted a royal audience, Van Langren traveled to

Spain in 1631, hoping to present his method for determining longitude and thereby gain further recognition at the highest levels. But being shy by nature, he was continually frustrated in his task because of his inability to find favor with any of the king's immediate advisors. He remained in Madrid until 1634, when the king ordered him back to Belgium without committing any interest to his ideas. The year was a fateful one for Spain: Isabelle died and rumors flew thick of France's imminent entry into the war, a move that would mark the end of Spain's hopes for victory. Philip, however, appears to have been sufficiently impressed by Van Langren's relationship with his sister to finance (at a low level) his continued efforts and to allow him the new title "Cosmographe du Roi."

The death of Isabelle struck Van Langren a hard blow. His work on the lunar map, which by this time seems to have included several early drawings (perhaps as many as thirty or more), fell into a period of despondency. A friend and colleague, Erycius Puteanus, replaced the archduchess as his immediate patron, encouraging him in various projects during the 1640s. Sometime between 1643 and 1645, Van Langren learned of other efforts to map and name the Moon. This news and possibly the memory of Isabelle's faith in the project urged him to quickly assemble and publish his own definitive version, thereby laying claim to a long held and reawakened ambition. By February 1645, Van Langren had finished an early manuscript version of his map, complete with nomenclature, and had applied to the king's Privy Council for permission to publish it. In his request, he wrote that he had "adorned [the Moon] with the names of eminent persons, as His Majesty has wished." We might therefore wonder if Van Langren's naming scheme was less chosen than commanded. The true body of the letter, however, addresses a different point:

> As the supplicant fears that some other person might well change the mentioned denominations and by such means alter and place in confusion the observations here achieved, and moreover prevent the supplicant from attaining his rightful compensation for such work, he hereby humbly requests it please your Majesty to expressly order his subjects to change nothing on said figure . . . that he issue a decree regarding this measure and privilege to be sent with all possible speed to a certain individual, in order to prevent him from

advancing further with his own design, to the disadvantage of His Majesty.[7]

This "certain individual" appears to have been Juan Caramuel Lobkowitz, the Polish astronomer whom Gassendi counted as a valued colleague and with whom Van Langren himself corresponded at the time.[8] In the above quotation, Van Langren calls on the king to secure by royal fiat a new realm of scientific discourse.

Such an appeal was hardly new. Galileo had named the four moons of Jupiter the "Medici stars" in the hope of securing the patronage of that powerful family and had even insisted that God himself had guided the choice. Here again, Galileo was very much a forerunner of what was to come. The power to name for all time and all humanity something new in the universe, something distant yet ever present, Galileo employed for the sake of momentary needs that were emblematic of the larger circumstance (dependence) of science at the time. The way he justifies and represents this choice in his preface to *Sidereus nuncius* may be the most eloquent expression of the mixed dependencies of scientists at that time and of their desire for fame and flattery among their patrons. In his opening "letter" to the "Most Serene Cosimo II de'Medici," Galileo states how all ages have tried "to preserve from oblivion and ruin" the names of "excellent men" by carving statues, building pyramids, erecting cities, and writing great works. Then he says,

> But why do I mention these things as though human ingenuity, content with these [earthly] realms, has not dared to proceed beyond them? Indeed, looking far ahead, and knowing full well that all human monuments perish in the end . . . human ingenuity contrived more incorruptible symbols against which voracious time and envious old age can lay no claim. And thus, moving to the heavens, it assigned to the familiar and eternal orbs of the most brilliant stars the names of those who . . . were judged worthy to enjoy with the stars an eternal life. . . . This especially noble and admirable invention of human sagacity, however, has been out of use for many generations. . . . But now, Most Serene Prince . . . behold, therefore, four stars reserved for your illustrious name, and not of the common sort and multitude of the less notable fixed stars, but of the illustrious order of wandering stars, which, indeed, make their

journeys and orbits with a marvelous speed around the star of Jupiter, the most noble of them all, with mutually different motions, like children of the same family, while meanwhile all together, in mutual harmony, complete their great revolutions every twelve years about the center of the world, that is, about the Sun itself. Indeed, it appears that the maker of the stars himself, by clear arguments, admonished me to call these new planets by the illustrious name of Your Highness before all others. For as these stars, like the offspring worthy of Jupiter, never depart from his side except for the smallest distance, so who does not know the clemency, the gentleness of spirit, the agreeableness of manners, the splendor of the royal blood, the majesty in actions, and the breadth of authority and rule over others, all of which qualities find a domicile and exaltation for themselves in Your Highness?[9]

It is necessary to look beyond the rhetorical excesses (a required part of all such writing at the time, partly meant to demonstrate the author's mastery of Latin technique) and to focus on Galileo's craft in another sense. In a masterly stroke, he has used the conventional gestures of patron flattery to link Cosimo's acceptance of immortality (in the form of named worlds) to acceptance of the Copernican system. Moreover, Galileo slyly suggests that doing so would be to act in accord with God's own wishes. The final portion of this passage, which alludes to "worthy offspring" who "never depart" from the Jovian beneficence, appears to be the author's own plea for being counted among the fortunate favored by de'Medici. Whatever the case, we shouldn't ignore a coating of happy irony that hangs over everything: today we know these four moons, not by the title given them by their discoverer, but as the "Galilean satellites."

Galileo's request for royal protection and support contrasts with Van Langren's plea for authoritarian guarantees. The Belgian cosmographer, however, appealing to Philip's own sense of troubled majesty, received approval for safeguarding his map, and the general decree was issued. What is not known is whether it was sent to the intended "individual," a Protestant.

Between February and the end of May in 1645, Van Langren completed at least four maps of the lunar surface, two of which are identical.[10] These vary in quality and detail. The most elaborate and finalized version is that of figure 10.2, which bears as many as 325 names; it is this image of which two copies survive. The other maps have fewer names and lack the specific date, author signature, and long inscription at the bottom. Moreover, the nomenclature shifts between images. Van Langren may have been in a hurry and was trying out different approaches to creating a suitable map, but he also may have consciously and cleverly directed his eye to suit different audiences and needs. The version he sent to the Privy Council to gain first-level approval, for example, had to make an immediate and powerful effect: this is the only map he illuminated. According to Bosmans (1903), the colors are beautiful and highly artistic, with the lunar disk a pale yet vivid yellow, the surrounding page tinted blue, the "seas" a deep aquamarine, and the craters a violet brown with traces of gilding. This is the simplest of the maps, with the fewest and most generalized features and craters with the least relief. Such simplicity was obviously calculated to allow for the best effects of color and to avoid overwhelming unscientific viewers with detail.

Another map version, located in Strasbourg, is closely similar to the final published version but lacks several detailed features, names, and inscriptions, and thus seems to be a draft. The version of figure 10.2 most fully obeys the general conventions for a published map of the time. In the mid-seventeenth century, it was common to publish maps rich in both artistic and textual decor. This included quotations in each of the four corners, as well as titles above and below the image. Descriptions of various voyages and discoveries might be placed within the map to cover the blank spaces of uncharted areas. Small biographies, outlining a particular mapmaker's purpose, abilities, and experience, engraved in Latin, also frequently appeared in a lower portion of the image. That Van Langren obeyed most of these conventions in his final map indicates that he intended his work for a knowing, scholarly readership. Similarly, his earlier versions, which lack much of the noted textual inclusions, may have been aimed at less educated audiences.

However common these written features may have been on maps of the Earth, they helped produce a unique and remarkable document of the Moon. Two textual elements in particular are responsible for this: the long Latin inscription at the bottom of the image and the nomenclatural scheme itself. The contents of the inscription show it to be part confession, part personal history, part paranoia, and part propaganda—in short, an inadvertently honest portrayal of its author's intentions and weaknesses. It is, in fact, worth quoting at some length because it provides one of the best examples of how the private ambitions and failings of a particular scientist can become material for the most public of nomenclatures (i.e., that of the heavens).

These things, remaining until now unpublished despite their great utility for mankind, despite even their necessity, Michael-Florent van Langren, mathematician and cosmographer to the king, offers to the entire world.

The lunar globe is at once the most obvious of the stars and the least known. Its geographic description was undertaken by me with great care and effort for her most Serene Highness, Princess of Belgium, Isabelle-Claire-Eugenie, Infante of Spain, a task I have completed. The great love of this princess for the sciences urged her to command my assistance for her observations of the Moon; still more, she visited me in order that we might contemplate together the secrets of that star. She well understood their great importance. Charging me therefore with letters in her own hand, she sent me to Spain, there to abide by the all-powerful king Philippe IV, to whom I was to present my observations, to publish them in his name, and to therefore furnish a reliable astronomical method for determining longitudes and distances of terrestrial locations and, through these, correct enormous geographical errors. Marine navigation might also make use of them.

In all this, the great king took a keen interest. He had me often called to his side, in order to assist him in observing the sky and Moon with the aid of the telescope. He further allowed me to give this selenographic description or lunar geography the title By the Light of Philip of Austria and to make it known to the world under the auspices of his name. He also approved setting down the names of famous persons for the luminous and brilliant mountains and islands of the lunar globe. These will serve to distinguish [such features], and it will be possible henceforth to make use of them in observations and corrections of an astronomical, geographical, and hydrographical nature. In his letter of response to her most Serene Highness, Princess Isabelle, [the king] ordered that the required expenses be paid. Yet, upon my return from Spain, in 1634, this august heroine, whose goodness, justice, piety, and clemency were known to all the world, was returned to the Earth en route to the skies, to there witness these great marvels at close hand.

It pains me to recall what misfortune of time my work, already in progress, then suffered interruption and, deprived of its support, vacillation. It began little by little to be disclosed to others, to such a point where the danger arose of seeing another make off with it and publish it under his name. Finally, the excellent don Emmanuel de Moura y Cortereal . . . governor of the Belgian provinces and Burgundy, etc. intimate advisor on finance, was informed as to the uniqueness and utility of the work. He foresaw the eternal glory which would be showered upon the king and gave permission for the publication of this selenography. . . .

Encouraged by this decree (that it might be for the benefit of the general public!), we have begun with the publication of this Philippian Full Moon, adorned with proper names. These are of kings and princes (who reign at present in Europe, as the patrons, promoters, and supporters of the mathematical sciences), more still of ancients and moderns, eminent in this field, who have acquired commendation and glory by the excellent monuments of their genius. We will also publish a book in their honor. To our profound regret we have been unable up till now (but hope to correct this soon) to learn the names and achievements of those foreigners of excellence in the sciences, in order to inscribe them equally on our brilliant globe.

We possess already thirty images of the waxing and waning phases of the Moon. These appear incessantly. We show distinctly all the details of the lunar surface, such as the islands and summits of the mountains. Removed very often from the continents, they appear instantly during sunrise on the Moon and disappear suddenly during sunset. They provide the means which are best, and of an almost daily employ, for the determining of longitudes . . . Such are among the things that, by order of that most Serene Princess, we demonstrated, in 1631, to scholars very knowledgeable, illustrious, and renowned in this science: E. Puteanus and G. Wendelin; then, in Spain, by order of the king, J. della Faille and B. Petit . . .

To avoid, however, the confusion that would result in astronomical and geographic observation from changes that someone might effect on these denominations of the Moon, we have distributed a large number of copies of this engraving, without expense and very widely, to all those whom we venerate as being today the benefactors, defenders and patrons of these studies. We therefore dedicate this image of the Moon most humbly to the kings, the princes, and the noble amateurs of these sciences. We request of them to approve our nomenclature of names, to change nothing, and to agree in the greater part of what we offer them here. . . .

5th of the Ides of February, 1645, Bruxelles

Prohibited by decree of the king: any change in the names of this figure, under pain of indignation; any counterfeit copies of said image, under pain of confiscation and three florins fine.

Given this day, 3 March 1645[11]

This was Van Langren's epitaph to all of his ambitious plans for work on the Moon. A more striking contrast to Galileo's confident, rhetorically complex, and clever appeal to Cosimo de'Medici could hardly be imagined. The year after this map was published, Puteanus died. Van Langren, deprived of his friend and second patron, abandoned the lunar surface for good and sought more mundane and profitable subjects, especially of a military sort, thereby currying royal favor during the last, particularly difficult and intense stage of the war.

In this inscription, which he chose as a legacy of his vision for the Moon, Van Langren shows himself, as in all of his activities and writings, to be exceedingly dependent—financially but even more emotionally—on the favor of those whose beneficence he seeks. Such favor, it appears, was his lifeblood. And this lifeblood spilled itself across the surface of his map. Indeed, Van Langren's Moon was exactly that—a personal absorbent of gratitude, faith, and obsequiousness. Lobkowitz, his sometime correspondent and friend, once described him as "a very polite and gentle man, worthy of more happiness than he received."[12] Should it be any surprise, given the sentiments expressed in his text, that Van Langren's nomenclature would be a plea to preserve the remains of a Catholic Europe about to disappear? The title itself, claiming Austria as a terrain of Philip, is an expression of the Spanish

Hapsburg's move to join forces with the Austrian branch of the family to fight the Protestants in Germany, Holland, Sweden, and elsewhere.

To the lunar seas, Van Langren gave titles such as Oceanus Philippicus, Mare Austriacum, Mare Borbonicum, Mare de Popoli, Mare Eugenianum (for the princess), Mare Belgicum, and Fretum Catholicum. The intervening continents, or *terrae,* he denoted with names such as Honoris, Sapientiae, Dignitatis, Pacis, Virtutis, and (again) Philippicus, thus implying the plenary virtues that existed among the referents just mentioned. The major craters are named after Catholic kings and princes, both living and dead, complete with their numeral designations: Philip IV, Ludwig XIV, Innocent X, Alfonsus XX, Carolus I, Ferdinand III, Christian IV, Ladislaus IV, etc. Each of these names is followed by a specific royal title; e.g., Carolus I, Reg. Britt. (Charles I, King of England). A few other craters of somewhat smaller size are named after queens and princesses, along with their titles. Saints' names— Augustine, Bede, Dominic, Francis, Ignatius, etc.—are reserved for headlands and capes. The minor craters are given the names of famous scholars, mainly mathematicians and astronomers, both ancient and contemporary: Hipparchus, Aristarchus, Kepler, Copernicus, Galileo, and Van Langren's personal friends (Puteanus, Wendelinus, Lafailli, and himself, Langrenus).

The hierarchy of this scheme is obvious and was not lost on astronomers of the day. The effort to claim the Moon as Catholic territory, to make Pythagoras and Plutarch the semantic subjects of Philip and Isabelle, is only too clear. The scheme overtly placed "science" at the bottom of coined importance and fawningly bowed to the imperial ambitions of a few select nations. The king of Spain appears five times on Van Langren's Moon as its true colossus. The lunar surface, according to this scheme, did not belong to astronomy; astronomy, however, belonged entirely under royal power. Van Langren's map therefore inscribed a truth that Galileo's rhetorical cleverness superseded: namely, that contemporary science was often plagued by a vulnerability to the caprice of individual rulers, whose favor never came without strings.

In the months just before his map was published, Van Langren spoke of his choice of names with Puteanus. These letters echo Gassendi's own memorial impulse toward friends and colleagues, and the qualities of coming failure that they bear cast over them a veil of sadness, even pity. It is as if the scent of death were in the air. Puteanus, for example, now less than a year away from his own demise, expresses his gratitude that he might see himself "written in your Moon and destined to live there always with you, as we have been together here below, on our mortal Earth." He advises Van Langren, "[D]o not use many ancient names. . . . Use those of our contemporaries, in order that they gain such renown that they will then work to fulfill their reputation."[13] Here is the same wish for community in the heavens that Lobkowitz expressed to Gassendi in his "we shall all be there, together." One hears in this the ancient Pythagorean tradition of the Moon as a resting place for restless souls. There is more, however. Puteanus also agrees with using the names of living princes, kings, and royal families and recommends expanding the pantheon of scholars, even to include such non-Catholics as Huygens. Why? "It must be done," he says, "in a manner such that our enemies should have no pretext for producing a new map of the Moon, in their own image."[14] The war in Europe, which lasted through much of the lives of these men, was not simply fought on the Moon they inscribed; it was lost there as well.

11

Johannes Hevelius

A Moon of Higher Origins

LIFE AND BACKGROUND

*I*n the inscription to his lunar map, Van Langren wrote of his desire to include on the Moon "foreigners of excellence in the sciences." One of these, whom he honored with a crater, would prove to be his most ardent rival. This was Johannes Hevelius (1611–1689), Protestant merchant, city official of Danzig (Gdansk), and observational astronomer par excellence, who also embarked upon a mission to chart and name the lunar surface, and against whom, in the last years of his life, Van Langren bitterly complained for having ignored his own work. Such competition was endemic among those with a serious interest in selenography. Hevelius himself, otherwise a man of generous temperament, once said of Galileo that he "lacked a sufficiently good telescope, or he could not be sufficiently attentive to those observations of his, or, most likely, he was ignorant of the art of picturing and drawing."[1]

This criticism might give us pause, yet Hevelius was the first to produce not a single map but a true lunar atlas. He was the first to create an extensive and remarkable collection of drawings on the lunar phases at roughly two-night intervals, with each of them named and each of them displaying numerous surface features as they advanced into and retreated from the borders of visibility. Hevelius was also the first to produce an atlas of a foreign world that was entirely similar to the Earth atlases then popular throughout Europe. No one before, nor for a long time thereafter, attempted such a document. It was, along with its naming scheme, a magnificent achievement by any standard but destined by history to fail.

In his early successes, Hevelius seems to have been favored more by circumstance and skill than by brilliance and originality. He was the eldest son of a wealthy brewer of German ancestry, who, like other middle-class parents with recently acquired means, was ambitious for his children and thus provided an excellent education for Johannes before involving him in the family business. The financial success of the brewery, whose ownership eventually passed to him, allowed Hevelius complete freedom from the necessities and burdens of patronage so commonly endured by scientists of his day. His affluence also gave him the means to construct the largest and most sophisticated observatory in all of Europe, one to which scholars, diplomats, and royalty all made pilgrimages. Beyond this, Hevelius's education and interests gave him considerable training in several crucial areas: drawing, engraving, instrument making, lens grinding, and printing. Perhaps less a scholar than Galileo, he was far more an able craftsman of his own astronomical cosmos. His observatory, built in the early 1640s, was the envy of astronomers everywhere. It contained not only a wide range of devices of his own design and manufacture, including the largest telescopes then known, but also all of the equipment required for high-quality engraving, an "optical plant" for producing lenses, and a printing office complete with brass and wooden presses.[2] One can only marvel at his ability to control, with a minimum of intermediaries, the means of his own scientific production—from raw observation to published result. In some ways, it was this liberty from the ordinary social restraints on science that was the most unique aspect about Hevelius.

ASTRONOMER BY CONVERSION

Biographical sources relate that a profound intellectual and emotional influence on the young Johannes came from his first tutor and later friend, Peter

Krüger, a well-known mathematician and astronomer.[3] It was Krüger who recognized in Hevelius a great aptitude for observation and for learning in general and who essentially converted him to the pleasures and discipline of astronomy. He did more than this, however, because he encouraged his pupil to take full advantage of his visual gifts by studying drawing, engraving, lens grinding, and instrument making. Hevelius was a more than willing student in all these areas and, while still in his teens, began to help his tutor with his research. Krüger's eyesight was not up to the task of scanning the heavens for extended periods; the young Hevelius was solicited for a significant portion of this work. This job was even more significant than it appears: Krüger's basic philosophy was that observation, not theory, composed the true center of astronomy and should be pursued systematically with the best instruments available.[4] The advent of the telescope, Krüger said, along with improvements in the quadrant, sextant, and other visual tools, ensured that whoever diligently followed the example set by Tycho Brahe could advance astronomical science more than at any time in history.

While still an adolescent, Hevelius was swept into the heart of the new astronomy and even given a substantive role to play. The opportunity must have been, at the very least, inebriating. But its fruition was to be postponed. In June 1630, when he was nineteen, Johannes was sent abroad to study law at the university in Leyden. The next year he traveled to London, Paris, and Avignon, where he met many of the century's most famous scientists and astronomers, including Pierre Gassendi, with whom he established a lasting friendship. But before he could go to Italy, where he planned to visit Galileo and other astronomers, Hevelius was called home by his parents to study law and jurisprudence at Danzig and thereafter to take his place in his father's business and in the administration of his native town, both of which were duties incumbent upon a prominent merchant's son. Having obeyed the wishes of his parents and neglected astronomy for several years (to his own private chagrin), he was called to the bedside of his former mentor, who now lay dying. Hevelius himself is the one to tell us of this, in the preface to his work *Machinae coelestis pars prior* (1673). He also informs us that, with his dying words, Krüger urges his favorite pupil not to abandon the subject that had bound them so closely in life. He quotes his friend and tutor's last words as these:

Because a rare eclipse of the Sun will soon be visible here, it is of great concern to me that I observe it [for the good of] astronomy. I feel, however, that it is not God's will that I be able to do this myself. Therefore, I urgently recommend to you, because I must part from this life, that you take over this observation, and that you dedicate yourself more actively to astronomy. I am convinced that you would never regret this, because the study of astronomy will not only bring honor to you, but . . . fame to Danzig.[5]

Krüger's words and imminent death worked a conversion upon Hevelius. The eclipse would be observed and recorded; astronomy would gain a new son. A sense of devoting himself, his wealth, and his position to something higher and more lasting took hold of the young Hevelius. Yet he was also a practical man with a family and serious responsibilities in business and government. He also seems to have understood that the doing of science included its own financial burdens. During the next few years he continued in his capacities as head of the brewery and city official but gave all of his free time and no small expense to building instruments; observing the planets, the Sun, and especially the Moon; and starting up a prodigious correspondence with other astronomers, scholars, diplomats, and officials throughout Europe. From this web of contacts, Hevelius was able to keep abreast of nearly all important news concerning the astronomical profession, including information about other observational efforts like his own. He was able to use this network, in turn, to transmit throughout Europe the progress of his own work.

SELENOGRAPHIA: A MAJOR FIRST WORK

For his first major project, Hevelius chose to observe and measure lunar eclipses. Finding all existing portraits of the surface inexact, he decided to produce a lunar map that would set a new standard for precision and usefulness. The telescope was his instrument; Galileo had proven it to be the tool of discovery, and Krüger had concurred. Hevelius found that existing telescopes were also insufficient for his purpose, so he constructed two of his own design, a six- and a twelve-foot version with magnification above forty times. Not long after this, he learned by letter that his friend, Gassendi, was engaged in an identical project to map the lunar surface. Reports about Hevelius's response to this news are varied; some have contended that he "wanted to renounce his own work; and it was only the pleading of Gassendi, and the assurance that he had abandoned his intention, that reconvened his [Hevelius's] own study of the Moon."[6] This seems exaggerated. For example, it is known that Hevelius wrote to Gassendi to inform him of his own project and to show him several of his finished drawings and engravings. Gassendi's response is telling:

> I owe you great thanks for the beautiful drawings . . . and am pleased that you possess such a fine telescope. You are gifted with such superior eyes . . . and are so clever at drawing that the representation could not be improved upon. Therefore, I not only agree, but urge you as much as I can, that you bring to light the description of the Moon that I had planned. For, while I, without any talent for drawing, must rely on someone else, you, with your rare gifts can not only draw the objects, but, what is even more important, can engrave the copper plates yourself.[7]

It is obvious that Gassendi was overwhelmed by the skill and self-sufficiency of Hevelius and by the promise that the latter's efforts would reach a far more speedy completion in published form. Gassendi, in fact, appears to have opposed publication of Claude Mellan's images in his own country in order to send them to Danzig in 1644, thereby effectively removing all known competition and leaving to Hevelius the honor of the first lunar map. Seeing two of these engravings, Hevelius resolved to broaden his own work considerably, from a single map of the full Moon to a full atlas of all of its phases. On this he appears to have labored night and day for a year or more, observing after sunset and engraving in daylight. Publication of Van Langren's map in May 1645 urged him to expand the project still further to include more than forty detailed drawings—several of the full Moon, others showing his instruments, and a number of optical sketches—plus nearly 500 pages of explanatory text with a glossary of 275 named features on the lunar surface.

The result, entitled *Selenographia: sive, Lunae Descriptio* (Selenography, or a Description of the Moon), is truly one of the grandest publications in all of seventeenth-century European astronomy. Pope Innocent X, upon being shown a copy of the work, remarked that "such a book would have no equal were it not written by a heretic." (Hevelius was a Protestant.) Appearing in 1647, its title was apt: it attempted nothing less than the most complete description in text and image of the lunar face and nearly all other telescopic phenomena of the day. True to Krüger's maxim, it did not venture deeply into theoretical, mathematical, or computistic territory. The first five chapters deal with lenses and lens making, the construction of telescopes, a very general history of planetary and stellar observation since classical times, and basic questions concerning the Sun and sunspots. These chapters provide a guide to "the practice of [telescopic] observation" itself.[8]

HEVELIUS AND THE TELESCOPE: AN INSTRUMENT OF DISCOVERY

It is a cliché of astronomical history that Tycho Brahe was the last observer of the heavens to rely solely on the naked eye. This is true but only in part. In a field renowned during the seventeenth century for so many "firsts," Hevelius fits this "last" more than anyone. The reason Hevelius decided against using the telescope for his work subsequent to *Selenographia*—that is, for the majority of his career, especially for his detailed observations of stars—was, in the beginning, a simple and straightforward one: it was not yet possible to make positional measurements with existing versions of the telescope. This capability had to wait for the introduction of micrometers (cross hairs) into the focal plane of the objective. Such an idea was stumbled upon, somewhat poetically it seems, by the English engineer William Gascoigne in 1640. In his letters, Gascoigne reveals that the concept came to him one day, "when I was with two convexes [lenses] trying experiments about the Sun,"[9] and he noted that one lens had been crossed by the strands of a spider's web. Unfortunately, Gascoigne was killed at the battle of Marston Moor in 1644 while fighting for the royalist cause, and his insight remained unknown. It was apparently repeated by the lens maker Eustachio Divini (1610–1685), who published a full-Moon map in 1649 with an inscription indicating his use of "fine threads" for correctly locating the "lunar spots."[10] The insight, however, appears to have languished again. A decade later, Christian Huygens (Père) discussed how he placed brass plates of various widths into the focal plane in order to measure the angular diameters of the planets (*Systema saturnium*, 1659). Yet it was not until the late 1660s, nearly thirty years after Gascoigne's original insight, that two French engineers, Auzout and Picard, produced a true caliper system for making precise stellar measurements. A few years later, micrometer-equipped telescopes became standard.

Until this occurred, the telescope remained what it had been in the hands of Harriot and Galileo: an instrument of discovery, not of measurement.[11] "Discovery," as idea, aesthetic opportunity, and practical ladder to fame, was one of the unifying themes of seventeenth-century European astronomy. It hung about the Moon, a "New World," like a brilliant halo. Galileo had by no means exhausted the visual treasures of the Moon; he had merely opened the chest. From then on, observers could themselves "discover" individual gems of detail that lay close by. Moreover, better quality lenses, longer tubes, and other improvements increased the exploratory power of the telescope. During its first half century of use, the telescope enthroned, stimulated, and at times, gave free rein to the observational side of Western astronomy, a side that was to decline thereafter under the aegis of "precision" and "measurement." Hevelius did the majority of his life's work between 1640 and 1670 (he died in 1687), exactly the period preceding the introduction of micrometers.

After publication of *Selenographia,* Hevelius abandoned the telescope for his own work of measurement. He employed large sextants and a six-foot azimuthal quadrant, willed to him by his beloved tutor, for measuring the positions of the stars. The accuracy of his eye proved itself the equal or better of micrometer-equipped sextant telescopes when they did arrive. This became evident during a celebrated and heated controversy in the mid-1670s, started by the often cantankerous Robert Hooke, who published a work that sought to cast doubt on the accuracy of Hevelius's results.[12] The young Edmund Halley was dispatched to Danzig in 1679 by the Royal Society of London to make a final determination and concluded his evaluations by vindicating Hevelius and his instruments.[13] Fortunately blessed with excellent vision, Hevelius could resist the changes then occurring and remain a *grand homme* of an earlier age. By the 1670s, the equipment in his observatory represented sophisticated versions of implements that were rapidly becoming outmoded elsewhere in Europe. It is in this sense that Hevelius was the last of the great naked-eye observers; as the frontispiece to his *Prodromus astronomiae* (which we will shortly discuss) strongly suggests, he knew it. His desire was to follow his tutor's wishes: to enter the ranks of the greatest observers and carry forward the world of classical observational astronomy.

Selenographia does not begin with a normal prose address to the reader expressing rhetorical humility or hope for the reader's elevation. It opens instead with a series of honorary poems on Hevelius himself, written in high style by various friends who were scholars, teachers, officials, or ministers. The tone is distinct in its courtliness: Hevelius is the muse of the Moon; his virtues surpass those of Columbus, Daedalus, and the Greek heroes; he has performed a total eclipse of Galileo; his name shall live forever, beyond time, etc. It seems to be a dedication more appropriate to a patron prince, yet it was commissioned and published by the author himself. The kings and nobles who came to visit the Hevelian observatory and who helped finance its rebuilding after a tragic fire may have recognized in its builder something of power and reputation akin to their own. As the Pope's words suggest, Hevelius's achievement had a political dimension within the context of the religious wars of the time—less overt than Van Langren's but still constituting a "claim" in which those opposed to Hapsburg Spain and Austria could take some degree of pride.

THE IMAGERY OF *SELENOGRAPHIA*: THE EYE SPEAKS

The true core of *Selenographia* is in its images. The diversity, careful execution, and aesthetic appeal of these pictures were primary reasons that this volume remained a standard text for the next 150 years. Hevelius's drawings bring together several traditions of lunar image making: the geometric tradition represented by Plinian-type diagrams of the solar system and zodiac; phase diagrams (e.g., of Venus and the Moon) drawn in simple black and white form, as in medieval texts such as Sacrobosco's *De sphaera;* epicycle diagrams, as in *Almagest;* and of course, Galilean-type pictures. In his drawings of the stars (e.g., figs. D and E in the original text), Hevelius offers decorative, artistic images, depicting not simple six-pointed features (as did Galileo) but complex, gem- or snowflake-like designs. Equally elegant is his use of allegorical *putti* (cherubs, representing divine agency), shown holding banners and demonstrating the use of instruments in the corners of two full-

Figure 11.1. Engraving from Selenographia *(1647) by Johannes Hevelius, illustrating the telescope and its various parts. Courtesy of Harcourt Brace & Company.*

Moon drawings (figs. P and Q in the original; shown in figs. 11.4 and 11.6 herein).[14] The presence of these framing cherubs has a special purpose: it tells us that the Moon is not merely an object of observation but also an *objet d'art,* a magnificent aesthetic phenomenon deserving of the highest skill.

Hevelius also included a detailed discussion of his instruments, complete with illustrations. These pictures, however, are more extraordinary than they might first appear to be. Each illustration consists of two parts: an upper portion displaying the apparatus fully assembled and often in operation; and a lower portion showing it completely dismantled, with the individual elements arranged so that the eye can see every surface (fig. 11.1). Letter designations are given for each part in both assembled and disassembled versions. In some cases (as in fig. B of a periscope), three stages of assembly are shown. The effect is a remarkable feat of artistic engineering:

the viewer himself is urged to put the instrument together with his eyes, piece by piece, step-by-step, and thereby come to an understanding of its exact use and operation.

So numerous are the engravings in *Selenographia,* so rich in variety and detail, that there seems to be a reversal in the overall relationship between text and image: the writing often seems illustrative of the pictures. Hevelius, however, is not the author Galileo is; his style is sometimes redundant, run-on, even indulgent compared with the Tuscan's elegant brevity. Far more care, it appears, was taken in the stylistics of the images in *Selenographia,* which seem to strive for some unique merger among naturalism, cartography, and baroque decorativism (figs. 11.2, 11.4, and 11.5). The title page to *Selenographia* (fig. 11.3), unlike that of Galileo's *Sidereus nuncius,* is replete with terms such as *accurata, delineatio, observationes, figuris accuratisime aeri incisis,* and *experiendi.* The claim is for a superior kind of eye and vision. It appears to be less intended to make the reader a virtual witness than to overwhelm him with the evidence of Hevelius's own powers of witness and documentation. Visually the pictures in his work invite one into a masterly world of seeing, constructing (instruments), and drawing. As the title page says, "[Q]uestions of astronomical, optical, and physical importance are both propounded and resolved" *(astronomicae, opticae, physicaeque quastiones pro-ponuntur atque resolvuntur),* with emphasis, we might suspect, on the final claim. In contrast to Van Langren, Hevelius did not draw the Moon as an overtly political idea; he did not wish to celebrate a royal patron but to make visible for other astronomers his own royal powers of observation. His true ambition was to enter the ranks of the greatest celestial seers of the past: Hipparchus, Ptolemy, al-Battani, Ulugh Beg, Copernicus, and Tycho Brahe.

This ambition appears to be inscribed on the frontispiece to his final work, *Prodromus astronomiae* (Astronomical catalogue), which was published posthumously. In this image, ten noted observational astronomers of the past (including all those just mentioned) stand together as a council of sages surrounding Urania, muse of astronomy, to whom Hevelius presents his work for judgment. The image is a minor masterpiece of baroque engraving, with its myriad details of clouds, cherubs, cartouches, animals, and unfolding banners—a kind of vision or *somnium* whose center is Hevelius's own appeal to join the pantheon of the canonical eyewitnesses of the skies.

The frontispiece to *Selenographia,* on the other hand, is no less rich in

Figure 11.2. *Full-Moon image from* Selenographia, *engraved by Hevelius and showing highly decorative style. Courtesy of Harcourt Brace & Company.*

meaningful iconography (fig. 11.3).[15] At the top, Contemplation, telescope in hand, parts the clouds of ignorance, through which shine the Moon (complete with surface features) on the left and the Sun (complete with sunspots) on the right. Immediately below, two *putti* hold up a scroll on which is written, "Lift up your eyes to the heights and see who has created these things" *(Attollite in sublime oculos vestros, et videte qui creaverit ista)*. The words are from *Isaiah* 40:26; together with the image, they seem to say that the true astronomer does not merely observe but also contemplates, reflects upon, and finally worships the heavens. Only in this way can he become the messenger of divine truth. Below the two *putti* stand two figures on pedestals holding a banner with the title of the work. On the left stands al-Haytham (here called

Figure 11.3. Frontispiece from Selenographia. *Courtesy of Harcourt Brace & Company.*

"al-Hasen" by his Latin name), the famous eleventh-century polymath whose treatise on the lunar markings was discussed in detail in chapter 5. On the right is Galileo, with a telescope in one hand, the title banner in the other, and fingers pointing noticeably downward at Hevelius's name. This may have been the author's way of acknowledging his profound debt to the Tuscan, whose own work on the Moon clearly served as a model. Beneath al-Haytham in a framed niche is the pate of a head underwritten by the word *ratione* ("reason"); Galileo stands above a niche inscribed with a human eye and the word *sensu* ("perception"). The overall message is therefore revealed: the powers of Arabic and European astronomy, though great in achievement, are nonetheless limited in separation and must be combined in any true selenography worthy of the name. Mind and eye must be joined in the hand, and the promise is that Hevelius has accomplished this.

This leads us back to the images themselves. In these the human eye has a continual, even unremitting presence. Beyond its mention and appearance on the frontispiece, it occurs in disembodied form in several other engravings, for example, in figure C illustrating the use of a periscope and figure H showing new stars as projected onto the zodiac by Rheticus and Gassendi. The eye is in residence as well in figure F (fig. 11.1), where an observer gazes through a mounted telescope; in figure G, where three different appearances of Saturn are shown; and in figure L, where an image of the Sun is being projected onto a screen for observation. Why this relentless emphasis on the power of human sight? For Hevelius, the eye is itself the ultimate tool of analysis and understanding. The eye produces knowledge even as the mind digests and reconstitutes it. Whether borne aloft by the telescope or guided with precision by a quadrant or sextant, the eye represents the first and final instrument of discovery.

THE MOON IMAGES: A BRIEF CRITIQUE

The engravings of the Moon's phases done by Hevelius for *Selenographia* set a basic standard for lunar image making that was not exceeded for more than a century. Not until the publication of Tobias Mayer's map in 1775 and its expansion into a lunar atlas by Johannes Schröter in 1791 *(Selenotopografisches Fragmenten zur genauern Kenntniss der Mondfläche)* were Hevelius's pictures finally given shelf space in the library of "history." Before then, no other images were so often copied, imitated, plagiarized, distorted, and referred to as these.

One way of looking at the total collection of lunar images in this work is to notice that they, like the pictures of instruments, provide an assembled-

Figure 11.4. Composite image of the Moon from Selenographia *(fig. P), depicting features visible within the total limits of libration. Courtesy of Harcourt Brace & Company.*

disassembled scene. Hevelius's lunar pictures present their subject in so many phases that they offer the reader (who is as much a viewer) a time-lapse succession of the full lunar cycle. One can, in fact, flip the pages to gain something approaching this effect. The viewer is taken through all forty phases of the lunar cycle, then presented with a composite image showing the limits of libration (fig. P in the original; fig. 11.4 herein); a cartographic map with named features and designated mountains, rivers, seas, etc. (fig. Q; fig. 11.6); and a final topographic rendering (fig. R; fig. 11.5) that essentially interprets the composite image (P) in light of the given map (Q). This is not all, however. Hevelius goes further and shows us sketch-type outlines, without shading or embellishment, of all of the lunar features he has observed from the various phases and placed within the lunar disk (fig. T). The author thus provides a tour of his own witness and a record of that witness. There is an entire "world" or "cosmos" of documentation presented here, one that has been assembled for us.

What about the quality of these images? Inferior in naturalism to those of Mellan, they are decidedly more artistic in execution and design. Their very artistry, however, suggests their limitations. The full-Moon image of figure 11.4 is flat, without any attempt at three dimensionality, and the same is largely true of the final topographical image of figure 11.5. Moreover, when we look closely at figure 11.5 or follow the progress of the phase drawings, we begin to notice a number of strangely regular, patterned embellishments or inventions. One such feature resembling a pair of glasses or a figure eight, is evident immediately to the left of the two *putti* in the upper right corner. Below this feature, roughly halfway down the right side (just left of the Mare Crisium), we see two little boxlike formations of unknown meaning. Closer to the bottom along this same side, to the left of the telescope held by the cherub, are two craters with curious clawlike markings. (In the phase diagrams, these appear with four such markings, giving a sand-dollar-like appearance.) Looking at the craters themselves, particularly those in the lower center, we can discern both linear and arclike patterns in their placement and a tendency to group like sizes together. The more closely we examine figure 11.5, the more regularity we perceive and, at the same time, the less precision in the rendering of specific features. It seems that Hevelius adapted the lunar face to a certain sense of artistic design and order. This does not mean that he falsified or sacrificed his perceptions. On the contrary, these images seem to express Hevelius's own search for visual meaning in what he perceived in the lens. The complexity of what he saw with his superior telescope and his superlative eyes was probably vastly beyond anything even hinted at by Galileo's pictures and text. Mellan's images also would not have helped him. Furthermore, he was not the artist Mellan was; the most complicated images in *Selenographia,* including the frontispiece, were drawn and engraved by others.

Galileo's pictures gain much by their simplicity. Hevelius's images seem to lose accuracy because of their complexity. Galileo had a specific meaning in mind: his engravings were intended to prove the rugged, Earth-like nature of the Moon. By Hevelius's day, this was established truth. The author of *Selenographia* did not have an option to use the same style, although, as we shall see, he sought a way around this through his naming scheme. His

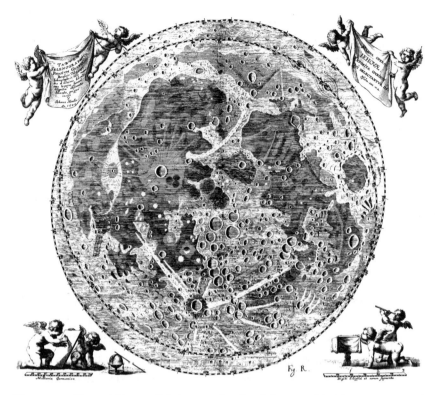

Figure 11.5. Topographic map of the lunar surface from Selenographia *(fig. R). Courtesy of Harcourt Brace & Company.*

images do not convince us of anything specific about the nature of the lunar surface. Here and there they evoke volcanic landforms and thereby seem to embody a particular theory of lunar origins. Yet these forms are developed only locally in one or two cases and seem countered by the range of other features bearing no affinity with volcanic formations. In the end, Hevelius's images have no clear pictorial rhetoric. They appear to have fallen back on techniques that render them as much designs as attempted likenesses.

HEVELIUS'S NOMENCLATURE: A HIGHER MOON

Hevelius produced a true map of the Moon, whose image is shown in figure 11.6. On this drawing, the Earth-Moon analogy is taken to a new extreme, showing seas, mountains, rivers, and lakes exactly as they were represented on terrestrial maps. Loyalty to the Earth is revealed even more in the naming scheme Hevelius chose, a scheme that also expresses his attachment to an earlier era of observation. In naming the Moon, Hevelius took an elevated pose. The words that explain his choices are in stark contrast to Van Langren's lament:

> Therefore, as the geography of the whole world, with all its very ample descriptions and commentaries, would be useful to no one without the names of regions, cities, towns, seas, rivers, mountains, and valleys . . . so it is in our case with selenography. . . . [All] and sundry comprised in the ambit of the World Machine (*Machine Coelestis*), whether animate or inanimate, from their initial discovery have been immediately given their own name. This problem was of concern to God himself, the creator of the universe, and to his handywork, Adam . . . [for] without this having happened, this visible world would be otherwise, in some part a confused chaos, that is, not perceptible and incomprehensible to man.[16]

To name is thus to take up the very powers of creation, to fertilize the world with its own visibility. "[T]he nature of our work," says Hevelius, "necessarily makes inroads on the prerogatives of God" (p. 224). It is thus a "dangerous path" that the author claims to have chosen, dangerous yet exalted. What, then, of his chosen scheme for naming the lunar face? To this question he provides an extended answer:

> Initially two methods offered themselves. . . . First, it would not be out of place to follow the example of the astronomers of old who named stars after men of surpassing virtue, people worthy of praise above others in the world (their intention was to establish a perpetual memorial for posterity). . . . I say for this reason, I had also proposed to myself to apply to the whole territory of the Moon the names of men of today who are famous and most learned, and especially those men who in our tempestuous times have come to greatly enlarge the studies of mathematics. . . . With this decision made, one would bring to this true celestial land the Ocean of Copernicus . . . ,

Sea of Kepler, Lake of Galileo, Marsh of Maestlin, Island of Scheiner, Peninsula of Gassendi . . . , Krüger Promontory, Strait of Eichstadt, Linemann's Desert, and so forth.

But this could scarcely be done without problems. . . . When I had given the business some thought, it seemed to me that. . . . I might seem to wish to curry favor by this method of naming and so I might perhaps inspire envy and hatred in my direction. For in this attempt, I realized that there are many places on the Moon's globe, and some are more eminent than others, and some are less notable. . . . And I thought it could happen that . . . for no good reason wrath could arise in some people because of what was done, because I might feel, due to religious creed, less kindly or honorably disposed toward someone . . . and perhaps I might have preferred one or the other [person] because of friendship or faith or some other less worthy reason. (pp. 224–225)

Note the direct swipe at Van Langren's own nomenclatural scheme. What the author says next, however, is more significant with regard to his general beliefs about the Moon:

Therefore I wished to do away with this means of naming the lunar spots and to choose some other way that is by all means safe, and as I really believe, *more fitting to our institutions and public usefulness.* Without doubt, the Moon could be named by the law of the antichthon [i.e. the Pythagorean idea of the Moon as a companion Earth], since in many respects it is nearly similar to our own Earth. . . . So I brought myself around to this very prospect [and considered] if there are such geographical names as are well known to apply them to the places on the Moon, provided that the hemisphere of the Moon facing us could be fittingly ordered to a certain part of the Earth's globe.

I immediately put my mind to this task . . . and had contemplated practically all of geography when I found to my perfect delight that a certain part of the terrestrial globe, and the places indicated therein, are very comparable with the visible face of the Moon and its regions . . . : namely, consider that part of Europe, Asia, and Africa that surrounds the Mediterranean Sea, Black Sea, and Caspian Sea, and all the other regions including and adjacent to them, which are: Italy, Greece, Anatolia, Palestine, Persia . . . etc.

Therefore, [I hit upon] the following naming system . . . : First, places should be considered so that the lunar spaces agree with the terrestrial ones; second, names to be chosen should be of the highest caliber and easily perceived by us, *especially those names used by the ancients, familiar to us and to historians.* For the names used in our times for terrestrial places greatly vary, and here and there a place is called this or that by different people. . . . In the meantime, no one needs to be persuaded that *the places on the Moon plainly correspond to our terrestrial places, even as regards form, size, and every proportion.* . . . I have used fictitious names but once or twice, and only in that place where there is no geographical [equivalent]. . . .

In this work itself I have distinguished with sure lines the shores of oceans, as is the custom in geography, and likewise marshes and mountains and valleys, so that even novices, at first glance, will be able to discern continents from marshes, ponds, and seas. . . . In this way, as we have said, adepts in astronomy will acquire quite easily the knowledge of all these things: since the names introduced here in their proper order will be *most familiar to historians, poets, and all those learned in letters.* (pp. 225–228, italics added)

Hevelius perceived in the Moon's surface, turned ninety degrees sideways (counterclockwise), an immediate resemblance to the classical world as known to and named by the Greeks and Romans. His nomenclatural image is therefore a completely terrestrial map, showing the lunar surface entirely reconstituted as the world of the ancients.

Hevelius has thus superseded Galileo in taking the Moon-as-another-Earth vision one step further. On the Hevelian Moon, we find the Mediterranean, Adriatica, Hyperboreum, Black, and Caspian Seas, all in their relative positions; islands such as Majorca, Malta, Sardinia, Sicily, Crete, and Cyprus; the Italian and Peloponnesian peninsulas; Africa, Libya, Arabia, and Armenia; the Atlas, Taurus, and Caucasus Mountains; the Alps; and any number of local features named for cities (Byzantium and Athens), classical gods and heroes (Letoa, Neptune, Hercules, and Cadmus), or specific mountains and valleys on Earth (Sinai, Olympus, Ida, and Hajalon). This list by no

Figure 11.6. *Lunar map with nomenclature from* Selenographia *(fig. Q). Courtesy of Harcourt Brace & Company.*

means exhausts the full range of Hevelius's scheme, but it gives some idea of its orientation. An alphabetical glossary of names includes 275 titles (several of which do not appear on the map) and describes each title in terms of its reference location on Earth and other names that have been attributed to it.

In his claim to have depicted the Moon's chief "seas, bays, islands, continents, peninsulas, capes, lakes, swamps, rivers, plains, mountains, and valleys," Hevelius has thus "discovered" through the power of names every type of Earth-bound geomorphic phenomenon. The craters, Hevelius says, are actually valleys and are therefore not round but only appear so because of their great distance, which obscures their irregularity. The nomenclature map (fig. 11.6) is drawn and hachured according to cartographic conventions of the time, with watery areas darkly outlined and mountains and ranges drawn as small rocky protrusions that cast shadows to the lower right (sunrise). Despite his own statements about being concerned with precision, Hevelius has drawn his map in a manner that is very difficult to use for locating most smaller scale lunar features. Due to the use of stylized hillocks, for example, many craters can not even be identified, and there are cases where a single name is used to identify several features. Scholars and historians of astronomy have often expressed dismay or confusion as to why Hevelius did this, or why, for instance, he drew the bright rays of Tycho and Kepler/Copernicus as long narrow lines of mountains.[17] Any such confusion vanishes, however, when one understands the degree to which he quite literally "saw" the Moon as another Earth. As a detailed example, he drew Sicily (Kepler on today's maps) as a large volcanic island and named its center Mt. Aetna, the most well-known and active volcano in Europe. His lunar Aetna approximates the general morphology of a strombolian-type volcano, with radiating ridges and a surrounding shield resulting from lava flows. In his text, Hevelius speaks of another feature he calls Mons Porphyrites (today, the formation Aristarchus) as being "without doubt" a volcano in "the midst of continual eruption," similar in its red color to "those known to us as Aetna, Heckla, Vesuvius, etc."[18]

Hevelius believed the Moon had inhabitants. His rationale, however, seems completely ironic: "Just because we do not perceive any beings there, it does not follow there are none. Would a man raised in a forest, in the midst of birds and quadrapeds, be able to form an idea of the ocean, and the animals which live there?"[19] Yet it is Hevelius himself, isolated in his magnificent self-sufficiency, who gave the Moon life. How could a world made in the image of a select portion of the Earth—a portion "highest" in civilization; most memorable to "poets, historians, and those learned in letters"; and complete with seas, plains, mountains, rivers, lakes, and volcanoes—not be inhabited? As figure 11.6 shows, even the Nile can be found on Hevelius's Moon, drawn with three streamlike lines ("inaccurately" debouching into the Bay of Sirte). Had Godwin a copy of Hevelius's map, his hero Gonsales might have explored the upper reaches of this lunar stream, perhaps even to its headwaters (the Mountains of the Earth?), centuries before its correlative was explored in terrestrial Africa. Yet the satires of Godwin and de Bergerac postulated such possibilities simply as a literary device. Hevelius, on the other hand, transformed them into a phenomenology. He stands at the true endpoint of that ancient vision of the lunar body, given voice in Orphic poetry (see chapter 3), which saw in this portion of the heavens "another world, immense, which the immortals call Selene . . . a world which has many mountains, many cities, many mansions."

Hevelius, no doubt, wanted his naming scheme to endure and saw the need to find some neutral, harmonious ground that might avoid the conflicts then raging throughout Europe. This neutral ground was a second charting of the classical world that might serve the unifying purpose of monumentalizing the (presumed) origins of Western scholarship. The Moon would be a surface on which every classically trained scientist, regardless of nationality or religion, could perceive his own origin. The effect, however, was problematic: where were the heroes of astronomy, past or present? The Hevelian Moon was a possession of European classicism but not of science, which was forbidden a single claim. In trying to create a vision for unity, Hevelius made invisible the very community he was trying to reach. His scheme also enthroned the classical world at a time when the ancients themselves were being increasingly rejected as models for understanding the physical universe. The seventeenth-century movement away from classicism in the sciences, beginning with the challenges to Aristotle, seems completely denied by the Hevelian nomenclature. The moderns who were then the body of science are nowhere to be found.

The fact that Hevelius did not include his own name on the lunar surface is a sign of the inevitable failure of his scheme. Whereas Van Langren aimed at the preservation of a Catholic Europe in the throes of defeat, Hevelius, with his adherence to ancient ways of seeing, sought to embody an ancient seat of origins whose means and thought were being superseded by those of the moderns. An eternal honorarium to Greece and Rome was not well suited to an age eager for new confidences of its own.

Riccioli

THE MOON AS A CONFLICTUAL COMMUNITY

A LUNAR LEGACY

*F*or nearly a century and a half after it was published, *Seleno-graphia* remained the standard lunar reference work in northern Europe, especially Germany, France, and England. Hevelius's maps and, to some degree, his nomenclature held sway, whereas those of Van Langren were forgotten or ignored. The reasons for this have a great deal to do with the impact of Hevelius's work on European astronomy, an impact that was greatly aided by his wide network of contacts and that extended well beyond the limits of lunar study. His influential volumes on comets, eclipses, stellar astronomy, and the constellations proved him to be qualified to enter the ranks of those he most admired, the great observers of the past. The care and style he demanded of the drawings in his works, which reached a zenith in his magnificent star catalogue of fifty-four maps (*Firmamentum sobiescianum*, 1688), established a standard that was difficult to exceed.

Despite such help, the Hevelian Moon did not survive the eighteenth century. Religious conflict helped ensure that even as early as the 1650s his Moon would fall from the sky in much of southern Europe, to be replaced with a new version produced by the Jesuit Father Jean-Baptiste Riccioli (1598–1671). It is, in fact, Riccioli's naming scheme that we use today. Religious strife, however, was not the only reason for Hevelius's waning Moon.

GIAMBATTISTA RICCIOLI: JESUIT INTELLECTUAL

Not a great deal is known about the life of the man who determined the future geography of the Moon.[1] Riccioli was a Jesuit priest and professor, first at Padua and later at Bologna, and a well-known, possibly even flamboyant intellectual of the day. He taught not only literature and theology, but also philosophy, rhetoric, mathematics, and astronomy. He was inducted into the Jesuit order in 1614 when he was sixteen years of age and spent the rest of his life as an author and teacher in the service of the Society.

At the time Riccioli joined the Society, the Jesuits were becoming directly involved in the Scientific Revolution through the efforts of thinkers such as Christoph Clavius, Giovanni Battista Mascolo, and Christoph Scheiner, the last of whom was especially involved in studies of the solar system. In addition to his famous work on sunspots, Scheiner produced a book comparing modern and ancient theories of the planets (*Disquisitiones mathematica,* 1614), which contained a crude map of the Moon with individual "spots" labeled by letters and described as "dark," "bright," "shadowy," and so forth. Scheiner's image is crude compared with Galileo's pictures, and much of his book is aimed at refuting Copernicus. Such was the basic stance of the Society throughout the century. By the time Riccioli began publishing his own astronomical work in the 1650s, however, this stance was becoming more tenuous and difficult to maintain, partly because of the competence, even brilliance, of Jesuit scientists themselves.

The standard science curriculum in Jesuit colleges, established at the famous school in Coimbra, Portugal, was based on Aristotle, Euclid, Sacrobosco, and related textbook commentaries by Jesuit professors. These professors were therefore heavily invested in pre-Galilean (mainly Aristotelian) cosmology. In their mission as the "schoolteachers of Europe and the New World," the Jesuits fought the new astronomy by carrying out and publishing their own work based on the latest technologies (e.g., the telescope) and experimental procedures. In addition to Scheiner, we see Jacques Grandami, in his *Nova demonstratio immobilitatis terrae ex virtute magnetica* (1645), drawing on William Gilbert's *De magnete* to "prove" the immobility of the Earth; and Charles Malapert attempting to refute Galileo and Copernicus through his observations on the comets of 1618 and stellar formations in the Southern Hemisphere (*Austriaca sidera heliocyclia astronomicis hypothesibus illigata . . . ,* 1633). The struggle for the heavens was integral to the Counter Reformation. Jesuit scientists considered themselves the intellectual defenders of the Church and, with knowledge and teaching as their weapons, attempted to match or even outdo the followers of heliocentrism in the detail, rigor, and scope of their work. During the second half of the seventeenth century, the Jesuits commanded great influence through the diverse writings of Riccioli and even more through those of Athanasius Kircher. Kircher produced a huge body of work in fields as diverse as medicine, botany, acoustics, geology, and mathematics. His major astronomical volume, *Iter exstaticum coeleste . . . ,* (1660), is a fascinating entry in the genre of the cosmic voyage. The hero of this story, subtly named Theodidactus, is taken skyward on a journey of the planets by the angel Cosmiel, who brings him first to the lunar surface so that he may see that "there are no plants, nor people, nor animals, nor any such living thing in the Moon" (*In Luna non sunt herbae, nec homines, nec animalia cuiuscunque speciei*).[2]

Like Kircher, Riccioli was a restless intellect in search of a center, a nexus on which he could focus his ambitions. During his own lifetime and for two centuries thereafter, he was renowned for his ability to consume vast amounts of reading material and to move between different subjects with an ease denied most of his contemporaries. This allowed or drove him to publish a work of even greater scope and magnitude than that of Hevelius. It is a work that sought nothing less than a comprehensive history and discussion of all the heavens, an ambition that is revealed by its title, *Almagestum novum* (The new Almagest).

ALMAGESTUM NOVUM: A STUDY IN IRONIES

To understand Riccioli's scheme of names—the scheme that eclipsed Van Langren and Hevelius and even the conflict between them—one needs to understand something of *Almagestum novum* and its context. The many difficulties, contradictions, and ironies of this book, as well as its magnificent scope and historical sweep, are a direct result of the Counter Reformation. It is evident from the title, for example, that the author adopted for himself the role of the Ptolemy of Catholic astronomy and geography. In fact, another of Riccioli's major projects (unfinished at his death, however) was to compose a modern counterpart to Ptolemy's *Cosmographia,* "a single great treatise that would embrace all the geographical knowledge of his time."[3] Whereas Ptolemy had originally sought to "save the phenomena" by reexplaining the classical cosmos on the basis of ecliptic orbits, Riccioli sought a parallel rescue by trying to uphold Aristotle and the biblical word. As one historian of the era describes it, "The Copernicans complained that the theologians, completely ignorant in mathematics, had hurled decrees against their system without reason. The desire to respond to them urged Riccioli to study astronomy; thus it was not strictly a love of truth, but rather the desire to plead the cause of the theologians and to defend the literal meaning of Scripture, that rendered our author an adversary of Galileo."[4]

The problem was that, as an able mathematician and observer, Riccioli soon found this position untenable. He attacks Galileo, Kepler, and Copernicus, but the manner in which he does so seems trivial. In his discussions on the various cosmological systems, for example, Riccioli finds none so elegant, accurate, or well conceived as that of Copernicus, yet he calls this "only a mathematical hypothesis." He then proposes a modified Tychonic scheme he knows is more complex and less plausible, in which Jupiter, Saturn, and the Sun revolve around the Earth; Mercury, Venus, and Mars orbit the Sun; and the Moon has only minor motion. "Doubtless he imagined this hypothesis in order to persuade the reader that he believed in the immobility of the Earth; yet, despite such efforts, one sees that without his cowl, [Riccioli] would have been a Copernican."[5]

Riccioli's true leanings are wonderfully revealed in the frontispiece to *Almagestum novum* (fig. 12.1), which appears to be modeled after Hevelius's engraving for *Selenographia* (fig. 11.3). It lacks, however, the latter's elegance and balance, being much more crowded, overdrawn, and visually difficult. Riccioli's image, like his book, is crammed with mythological, biblical, scientific, and classical allusions. In its upper portion, for example, the planets, the Sun, and a comet are all carried by separate cherubs. These scenes are arranged on both sides of a radiant divine hand whose three outstretched fingers are labeled with the three aspects of solidity given to the world by God: *numerus, mensura,* and *pondus* ("number," "measure or dimension," and "weight"). The Sun is drawn with a face, resembling the common representation in ancient Rome, whereas the Moon is given its modern, cratered surface. Below on the right stands Urania, holding a balance that weighs the system of Copernicus against that of Riccioli (Tycho Brahe), with the latter being the heavier one. "It will not be tipped for all time," says the muse of astronomy enigmatically *(Non inclinabitur in saeculum saeculi).* On the ground lie Ptolemy and his system, discarded; he speaks the words "I am raised up while I am corrected" *(erigor dum corrigor).*[6] Copernicus's system is the lighter, yet it is also the higher, closer to God and directly above Ptolemy, who seems to be looking upward at it, not at Riccioli's system.

Opposite Urania, who is covered with stars, stands the figure of a bearded man clothed only in a loin wrap, his body embedded with a constellation of eyes. He represents the classical figure of Argus Panoptes, the all-seeing and ever vigilant. The arm of this figure points upward at the divine hand; the words from *Psalms* 8:3 emerge from Argus's lips: "I look upon thy heavens, the work of thy fingers" *(Videbo cælos tuos, opera digitorum tuorum).*[7] There is irony in this unlikely merger of biblical and classical allusions. The most famous tale involving Argus, known to every intellectual of the late Middle Ages and Renaissance, was that of Io in Ovid's *Metamorphoses,* in which the many-eyed figure is sent by Hera to keep watch over her earthly rival. Taking pity on Io, whom he had seduced, Zeus sends Hermes to lull Argus to sleep, one eye at a time, and then to slice off the slumbering head.[8]

Riccioli, in the very content of his "science," personifies the confusions often suffered by the scholarly world during the Counter Reformation. Bound to Rome and to Roman Catholic traditions of teaching and learning, yet a thinker, observer, and experimenter of no small ability, Riccioli was propelled into a domain where the knowledge his faith decreed as the truth was in constant retreat, being progressively replaced in most European universities as outmoded and medieval. The Inquisition had forced Galileo to recant, and Riccioli seems to have agreed, refusing even to read his works. Yet his own skillful experiments proved and even refined Galileo's law of

Figure 12.1. Frontispiece from Almagestum novum *(1651) by Giambattista Riccioli. Courtesy of Owen Gingerich.*

falling bodies and led to a profound advance in using the pendulum to measure time, which Galileo had tried and failed to do. If the Tuscan had ever been brought near the flame, Riccioli would undoubtedly have felt the heat.

The narrative of *Almagestum novum* also reveals a magnificently complex and unstable work. This is due to its author's intellectual and spiritual librations, not only between science and theology but also between literature and astrology, classical myth and biblical tales, and history and biography. Much of *Almagestum novum,* in fact, is not a technical treatise but an encyclopedic compendium of earlier writings on astronomy, especially the Moon, in almost every conceivable genre of prose. For this reason alone, the book became a standard reference well into the nineteenth century. Yet it makes for a work of many voices often at odds with one another. Its beginning pages, for example, include all of the following: a straightforward historical outline of astronomers along with commentary on their work; the story of Prometheus abandoned to the Caucus in order to discern the laws of heaven; the "true" tale of a dead saint whose blood liquefies on the date of the autumn equinox; an elegant method for calculating the Earth's curvature; and a discussion of the Inquisition's lack of an official decree regarding the Earth's immobility ("however, as Catholics, we are obligated by prudence and obedience . . . to teach nothing which is absolutely contrary").

Riccioli wrote the work while serving as prefect of instruction at the University of Bologna and was granted leave from teaching in order to complete it. *Almagestum novum* was thus known and considered by the authorities as a highly significant effort to help stem the tide (no doubt with Hevelius at the crest) of rising heretical views on the heavens. Riccioli seems to have produced it in a fever pitch in only three years. Although it was used quite widely, it was also used variably: on the one hand, by Jesuit professors as far away as Scandinavia who were intent on disproving Copernicus; and on the other hand, by scholars who had long accepted the heliocentric view. Its greatest use was as a reference text for every survey of the physical and mathematical sciences written during the eighteenth and nineteenth centuries.[9] Yet authors in this category rarely fail to censure *Almagestum novum* for its rejection of the New Astronomy.[10] Praise for its erudition, qualified appreciation for its science, and condemnation for its basic philosophy: such was its larger reception.

Of all its efforts, the only one to survive interest down to the twentieth century is *Almagestum novum*'s lunar nomenclature. In a way, this seems apt, for much of the conflict in this vast work of seventeenth-century thought is embodied in the type of community its author placed on the lunar orb.

RICCIOLI'S NAMES FOR THE MOON

Almagestum novum appeared in 1651, the culmination of attempts at lunar mapmaking that followed in the wake of Van Langren. Most of these attempts showed the direct influence of Hevelius, especially his images P and R (see figs. 11.4 and 11.5), and did not include any new proposals for a lunar nomenclature. The general degree of accuracy and the style in the images of Anton Schyrlaeus, Eustachio Divini, or Gerolamo Sirsalis, for example, vary little. Their decorativeness is far more in line with Hevelius than with the naturalism of Mellan or Galileo. This is also true for Riccioli, but here the debt to Hevelius is openly admitted, and this acknowledgment marks it as unique.

In all of *Almagestum novum* there are very few images. The only pictures, in fact, are those of the Moon, which were drawn not by Riccioli but by his associate and fellow Jesuit, Francisco Maria Grimaldi, the well-known physicist. These images appear on two separate pages: one with a full-Moon drawing lacking nomenclature and surrounded by four smaller images of various phases; the other complete with names, as shown in figure 12.2. The first image, meant to display surface features only, is said to be "composed from the most frequently occurring phases selected in order to give as much detail as possible."[11] Riccioli notes further that "if one were to represent . . . all the possible features of the Moon, no fewer than three hundred drawings, not (as Langrenus believed) thirty, would be required, because the Moon almost never presents herself to us in exactly the same manner."[12] The author states that Van Langren had sent him several of his unpublished drawings. After comparing them with those of Hevelius, he decided that, although Van Langren had precedence, Hevelius was superior in every other way, including overall conception, drawing skill, and creation of a great work, *Selenographia*.

The first page of lunar engravings in *Almagestum novum* is also significant for directly acknowledging in its title the work of Van Langren, Hevelius, Divini, Sirsalis, and other lunar scholars of the 1640s, all of whose efforts are said to be "partly adapted, partly corrected." This type of citation, with its direct admittance of debt to others, was an extraordinary addition by Riccioli and Grimaldi. It had never appeared before on any lunar image. The need to seem original—however impossible in the wake of Galileo, Mellan, and their imitators—was a necessity for those seeking fame from the new discoveries and growing public interest in the Moon. Printers also wanted their books to sell well, and the semblance of originality was a premium with them, too. Any glance through the many images published between 1620 and 1700 would reveal how many were simply stolen or slightly modified from the work of others (especially Galileo and Hevelius). Riccioli and Grimaldi's decision to allude to their predecessors is singular and is reflected in the naming scheme used on the second lunar image.

This image is shown in figure 12.2. Grimaldi clearly sacrificed artistry to clarity of the names. The librational limits are copied from Hevelius, as are many of the features. There is some consensus among scholars today that the image is generally more accurate than those of Van Langren and Hevelius.[13] Despite this, as an image, Grimaldi's version of the Moon is closer to Van Langren's generalized map than to Hevelius's decorative design. It lacks the artistic interpretations and imposed regularities of the Hevelian versions. It does not pretend to be a true observation, but all of its names identify specific features. Riccioli states in his text that he has chosen elements of Van Langren's naming scheme over that of Hevelius. He has done this, he says, because the analogy used by the latter—the Mediterranean region as a nomenclatural model—is a poor one, without much direct correspondence of features. Moreover, Hevelius chose "names of our ancient geography, too little known today."[14] The Hevelian Moon was not only too small but also out-of-date, geographically speaking. There is irony here: having cast his intellectual allegiance to the old Scholastic view of the universe, Riccioli finds nothing odd in chastising a contemporary for being himself too much in league with the ancients.

The Riccioli/Grimaldi nomenclature was actually loyal to many masters. This made it historically very shrewd and even brightened it with humor. The scheme avoided Van Langren's Catholicism yet followed his and Gassendi's lead by reinhabiting the lunar craters with the names of famous astronomers—this time without restriction as to era or faith and with certain placements that had immediate significance to the politics of knowledge at

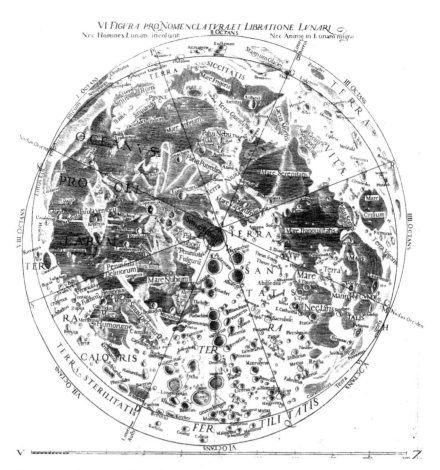

Figure 12.2. *Lunar map from* Almagestum novum, *showing the nomenclature of Riccioli and Grimaldi. Courtesy of Ewen Whitaker.*

the time. For this reason alone, it gained preference within a sizable portion of the scientific community, being favored by Hooke, Huygens, Mercator, and Wren, as well as most astronomers in southern Europe.

The scheme consisted of two basic parts, one devoted to major land and water forms, the other to craters. Seas, continents, lakes, and peninsulas were named after the effects and influences (especially meteorological ones) that had been attributed to the Moon by various (mainly Western) peoples and eras. The seas in particular show this in their invocation of fecundity (Mare Fecunditatis), serenity (Mare Serenitatis), madness (Mare Crisium), stormi- ness (Oceanus Procellarum), mist and vapors (Mare Humorum and Mare

Vaporum), rain (Mare Imbrium), and cold (Mare Frigoris). All of these names are still used today. Continents were similarly named, usually in opposition to their watery relatives: heat (Terra Caloris), sterility (Terra Sterilitatis), healthiness (Terra Sanitatis), liveliness (Terra Vitae), and cheerfulness (Terra Vigoris). These names, however, do not appear on current lunar maps.

It was in his naming of the craters where Riccioli was most inventive and where the issue of acknowledgment was most significant. These, he decided, would embody the progress of thought about the Moon. Riccioli divided the lunar surface into eight portions and conceived a hierarchy that would vertically distribute famous astronomers and philosophers according to era and subject matter across these octants. As discussed in *Almagestum novum,* those of ancient Greek ancestry appeared toward the top in octants I, II, and III (see fig. 12.2); those of Rome came just below, toward the center of the image, in octants IV, V, and VI. These were followed by medieval European and Arabic thinkers, placed toward the lower half of the image. These names did not include merely astronomers but also the authors and translators of important works that proved crucial to medieval learning, such as Martianus Capella, Bede, Alcuin, Geber, Alfraganus, and Dante. (Of these names, only the last was later removed.)

Below these ancients, in the bottom portions of octants V, VI, and VII and in the whole of octant VIII, the names of the moderns were placed, the most contemporary at the very outer edge. Included here were a great number of Riccioli's colleagues, and more importantly, a large number of his heliocentric adversaries. In a stroke, perhaps, of melded spite and wit, Riccioli cast the leaders of heliocentrism far out into the Sea of Storms (Oceanus Procellarum). Copernicus was kept closest, within sight of land. Kepler, however, was banished to a foreboding volcanic island (Insula Ventorum); Galileo was exiled still farther toward the outer margin, as distant from the ancients as possible. All three were left to a state of floating removal for having unjustly tried to make the Earth mobile. For readers who knew something about the lives, works, and personalities of this holy trinity of the new astronomy, Riccioli's choices had even deeper and more satirical meaning: Copernicus had been a local canon, remaining closest to the "true Church"; Kepler, because of his Protestant views, suffered near exile from his post of astronomer to the Holy Roman Emperor; and Galileo rarely acknowledged a debt to anyone. Hevelius, on the other hand, another heliocentrist, was granted a sizable crater on land just beyond Galileo in Terra Caloris. Van Langren was placed in a similar position but on the opposite side of the Moon, in the Sea of Fertility. This was actually a crater that Van Langren had named after himself on his own map. Riccioli was thus justifying his overall choice while framing it with a new, appreciative significance. Van Langren was granted the "fertility" of his vision, and Hevelius found himself at the margin of "true" astronomy.

Riccioli and Grimaldi did follow Van Langren in one minor respect. Clustered around the Sea of Nectar, they placed a number of holy saints (Catherine, Theophilus, Cyril, etc.). All of these, however, had some connection to astronomy; the relevant stories, sometimes of a mystical nature, appeared in the text of *Almagestum novum.* These names remain on current lunar maps, although stripped of the "St." designation. At the same time, lest this be mistaken for religious provincialism, it should be noted that even more Arabic and Jewish names appear just west of these saints, in the Land of Healthfulness (Terra Sanitatis). Here we can find the names of astronomers and mathematicians such as Azophi, Albategnius, Arzachel, Abenezra, and Alpetragius. We look in vain for al-Hazen (al Haytham), Averroes, Avicenna, and Albiruni, but no matter; these were all added later by others who were inspired by Riccioli and Grimaldi's ecumenical citation of the past. Such inclusion of Arabic astronomers is extremely significant from a cultural standpoint, as their names and works had continued to decline in influence in European universities from the fourteenth century onward.[15] The importance of Arabic astronomy to Europe was (and is), however, beyond all reckoning, since it was through Arabic translations, commentaries, and improvements that ancient science entered Europe.

In a profound reversal of a long historical trend aimed at erasing a portion of history, Riccioli put the Arabs back into the European sky, indeed into the very center of European astronomy, by inscribing them on the lunar face. He placed them not just anywhere but shoulder-to-shoulder with some of the most hallowed names in Western science: Ptolemy, Hipparchus, and Agrippa. One cannot help but wonder at Riccioli's audacity and his symbolic boldness—so reflective of his conflicting loyalties—in his "overthrowing" of Scholasticism by granting Albategnius a larger and more central crater than that given to Aristotle, who floats virtually alone in the chill waves of the Mare Frigoris! At the outermost edge of octant VIII, immediately below

Hevelius, Riccioli baptized two craters with his own name and that of Grimaldi. His own feature is smaller than that of his friend and collaborator. Indeed, it is the last feature to be seen along the entire western edge of the lunar face.

A FEW REFLECTIONS ON THE
RICCIOLI NOMENCLATURE

Riccioli's naming scheme calls upon a more varied astronomical culture than that of Van Langren and Hevelius combined. The names on his map are of persons with an enormous variety of connections to astronomy—scientific, astrological, literary, biblical, mythical, and philosophical—yet they also leave "science" preeminent. The Riccioli and Grimaldi map contains, in effect, a symphony of voices that largely define the Moon of Western culture—including its Eastern ingredients—down through the millennia. The title of the map *(Moon with Nomenclature and Libration)* carries beneath it the sentence: *Nec Homines Lunam incolunt, Nec Animae in Lunam migrant;* i.e., "Neither do humans inhabit the Moon, nor do spirits migrate there." It was Riccioli's feeling that Holy Scripture did not support any dwellers on the Moon, whether of flesh or spirit. Divested of literal life, the lunar surface could be peopled with the textual and historical community of which Riccioli considered himself a part, a collegium that his work sought to collect, teach, convert, and animate.

This encompassing quality was not complete, however. Riccioli left out a range of non-Western thinkers whose names he probably knew. Ulugh Beg, for example, built a huge observatory in Samarkand during the fifteenth century and compiled a highly valuable star catalog, which was popularized by Hevelius, who included Beg among his sages of astronomy. It is not true that the Riccioli/Grimaldi nomenclature succeeded merely because it fed the ambitions for fame of the community that had already gained possession of the Moon (in an epistemological sense). Its endorsements need to be seen in broader terms. The seas for example—today the most prominent and well-known features on the lunar surface—call upon common beliefs, many of which are truly ancient in origin, that existed in Asia and Africa as well as Europe. These beliefs relate the Moon to such things as agriculture and the harvest, fertility, birth, growth, the vicissitudes of weather, the crises of human psychic life, and the traditions of magic embodied in astrology. All of these notions cross many boundaries of class, education, culture, and time; no doubt they seemed wholly embracing to Riccioli. Moreover, Riccioli's vertical hierarchy of personal names cast a degree of irony (i.e., traditional modesty) over the promise of possible immortality: the lucky few who might one day join this list could expect nothing more than to appear at the bottom or near the outer limits of visibility. The layers of meaning involved in placing such controversial figures as Copernicus and Galileo in a region of "storms" or Aristotle in a frigid sea (Mare Frigoris) can not have escaped the notice or appreciation of Riccioli's contemporaries. In this sense, the map reads (and was no doubt read) as a code, a historical document merging private and public sensibilities, full of sectarian biases, and yet ripe with loyalties to the Moon as an object of Western cultural imagining. As a form of flattery inscribing a "textual community," this map had the ability to include a much larger membership than astronomers alone.

The Riccioli/Grimaldi nomenclature therefore recommended itself on a number of levels. Textually, it suited the age perfectly by both incorporating and exceeding the politics of science at the time. Though it included partisan elements, it also appealed to a "higher" and more harmonious realm: with its poetic, historical, and scientific qualities, it nearly personified the scholarly gentleman of the late seventeenth and eighteenth centuries. Its political content, though no less reflective of the period than that of its competitors, neither denied nor centralized the conflicts of the Counter Reformation but included them in a manner subtle enough to allow "astronomy" and "scholarship" to shine more brightly. Although the moderns are spatially given a low standing on this map, they are more numerous than the ancients—more numerous, in fact, than the ancients, medievals, and saints combined. Despite Riccioli's conservatism, his map documents the perception—dear to Galileo, Bacon, and the whole trend of seventeenth-century science—that more worthwhile work had been done in the last one hundred years than during the whole of the preceding millennia. In contrast to Hevelius, the Riccioli nomenclature incorporates the classical world but reveals it as limited, fallible, and superseded. The Moon here—our Moon today—finally becomes an embodiment of its own history in the West.

SUCCESS OF THE RICCIOLI/GRIMALDI MOON

The reasons why this naming scheme eventually gained favor over that of Hevelius are straightforward. As Ewen Whitaker has pointed out, the classical Mediterranean world of Hevelius's Moon yielded titles that were often unwieldy (e.g., Promontorium Freti Pontici) and were soon exhausted. Moreover, Hevelius had assigned single names to multiple features, thus creating an obstacle to any mapping at smaller scales.[16] Riccioli himself noted that the names of classical geography were becoming less known, a situation that would worsen during the following century with the final downfall of Latin as the language of European science.

Although the nomenclature of Hevelius was difficult, outdated, and self-exhausting even by the late seventeenth century, it did not disappear overnight, nor is the path of its eclipse obvious or predictable. Several crucial trajectories of influence, however, can be sketched. In Catholic France (and later in England), Hevelius was heavily favored for his images, but Riccioli's nomenclature soon dominated. For example, the famed French astronomer, Jean-Dominique Cassini, first director of the Paris Observatory, chose the Riccioli/Grimaldi naming scheme for one of his lunar maps at a crucial moment.[17] Between 1679 and 1680, Cassini had two artists, Leclerc and Patigny, engrave a large lunar map of excellent quality, imitative of the general style of Hevelius but containing much more detail and showing impressive topographic form. This map, one of the most famous in the entire history of selenography, carried no names, but a smaller version of it was produced in 1692, apparently for use in longitude determinations in the event of an eclipse. To this smaller version, Cassini appended a table listing Riccioli and Grimaldi's names for most of the features. A slightly reduced copy of this map was published in the annual astronomical review, *Connaissance des temps,* and continued to appear there throughout the eighteenth century.[18] This established the map as a standard, which was used in many publications on the Moon and planetary astronomy, not only for professional audiences but also for university teaching and popular writings. At the end of the century, the great French astronomer Lalande was full of praise for Hevelius's observations but utterly dismissive of his naming scheme, which he called "rather bizarre."[19]

In England, several factors seem to have determined the fate of Hevelius's nomenclature. It is likely that the cantankerous Hooke remained unfavorably disposed toward his former opponent in the telescope–vs.–naked-eye debate (see chapter 11). Hooke was not known for his affability or forgiveness, but he was renowned for his scientific work and leadership, first as curator and later as president of the Royal Society, which granted him much influence. A second source of favoritism for Riccioli and Grimaldi may have been Isaac Newton—if not directly, then through his most ardent popularizer, John Keill. Keill was a lecturer in the sciences at Oxford. In 1702 he published *Introductio ad veram astronomicam . . . ,* a work he later translated into English at the request of the Duchess of Chandos. Partly because of its skillful popularization and spirited defense of Newton in his priority dispute with Leibnitz, the English version was reprinted several times during the eighteenth century, well after Keill's own death in 1721. It appears to have been a standard reference for students and educated lay people as late as 1750 and was translated into several European languages, including French, German, and Dutch. The Dutch version, rendered by Johan Lulofs, a professor of astronomy and mathematics at the University of Leiden, was exported to Japan, where it became the foundational work of modern physics and physical nomenclature in that country.[20]

Keill's book contained a simplified version of Hevelius's map P (fig. 11.4) with the Riccioli/Grimaldi nomenclature imposed upon it. Keill added to it a new name, Flamsteedius, located in the southwest quadrant. This was apparently done while John Flamsteed (1646–1719), Astronomer Royal at Greenwich, was still alive. Flamsteed was embroiled in a long-standing dispute over his exacting observations of the Moon—measurements that Newton desperately needed and continually demanded in order to complete his lunar theory based on gravitation. In 1712, claiming Flamsteed's work was public property due to his official post, Newton and Halley published a large portion of his work. Flamsteed was incensed, and he tracked down and burned much of the unapproved edition. His own *Historia coelestis britannica,* containing both his lunar and extensive stellar observations, appeared in 1725, after his death and only two years before Newton's passing.[21] The specific addition—by one of Newton's most ardent champions—of Flamsteed's name to the canon of lunar features therefore takes on added significance. Was this move an attempt at reconciliation? Keill did not, after all, add

"Newtonius" to the Moon's nomenclature. Was it a sign of admitted defeat? Because of Keill's patriotic belief in public understanding of science, such interpretations seem less probable than the desire to add a recent British astronomer to the Riccioli/Grimaldi scheme. This scheme, after all, lacked any such inclusion: England was present on the lunar surface in the names of medieval thinkers such as Bede (Beda), Alcuinus, and Sacrobosco. This was in marked contrast to France, Germany, Spain, Italy, and Poland, which, in the Ricciolian nomenclature, were all amply represented by the moderns.

John Keill's example of augmenting Riccioli and Grimaldi's scheme with the names of other recent astronomers was followed by a later Jesuit scholar, Maximilian Hell (1720–1792). Hungarian by birth, Hell performed most of his work in Vienna and published in 1764 a set of lunar tables *(Praecepta pro usu tabularum lunarium)* along with a reduced copy of the Riccioli/Grimaldi map that bore several new names. The author retained Keill's "Flamsteedius" in the same area of the Moon and even added "Halleyius." Hell's map was not especially impressive or well known. It might easily have passed into oblivion were it not for its influence on the man who was to establish, once and for all, the choice of the Riccioli/Grimaldi scheme as the geography of the Moon.

This geography was established by the pathbreaking lunar atlas of Johann Schröter in 1791, which thereafter set the pattern for lunar maps down to the photographic era. Using a reflecting telescope vastly improved in resolving power over those of the previous century, Schröter identified, drew, named, and measured literally thousands of new features on the Moon's surface and did so in accordance with the Riccioli/Grimaldi scheme. Although it is beyond the scope of this book to discuss Schröter's life and work, it seems appropriate to quote from the portion of his work, *Selenotopografisches Fragmenten* (1791), in which he specifically talks about his nomenclatural decisions. This comes in Part I, Book 3, Section 32, where the author notes that he has observed at least 6,000 to 7,000 lunar features, "as many as the features identified in the heavens by our contemporary star charts." Riccioli listed no more than 244; thus it became obvious that a simple, complete, and relatively easy-to-remember nomenclature had to be chosen. To conceive a new system altogether, Schröter rightly says, would be "unreasonable and irresponsible" because it would cast "disadvantageous confusion" over previous observations.

In this respect, I have therefore retained without the slightest alteration the Ricciolian nomenclature as the most commonly transmitted on lunar charts, while simultaneously providing equivalents from Hevelius for those more accustomed to his system. . . . So that further progress in lunar topography will occur in the future, one will need to discuss and write about the [many features here identified by me for the first time], just as one did previously about such well-known locations as Tycho, Plato, Manilius, and the like. Indispensable for this purpose is an appropriately distinct and identifying naming scheme. I openly admit, however, that drawing up the present lunar maps according to this demand resulted in several difficulties. With such a great number of significant features, one must indeed, as with star charts, seek refuge in the use of letter designations from various languages. . . . Yet to identify everything new by means of letters would be too clumsy and place an unnecessary burden on the memory. . . . Therefore, to facilitate the powers of recollection I have followed the example of Father Hell and have designated the most pre-eminent locations and salient portions of the lunar spots with the names of famous astronomers and students of nature not found in Riccioli's nomenclature—indeed, *names that occurred to me without any relationship to proportions.*[22]

Schröter is entirely conscious of his potential role in the history of selenography. He attempts to successfully mediate between past and future, and this is precisely what he did. His massive atlas, far larger than either Hevelius's *Selenographia* or Riccioli's *Almagestum novum,* proved to be the determining work in the history of lunar nomenclature. It is interesting that, as a German, he notes that Riccioli's naming scheme is already the more preferred. Even in much of northern Europe, it would appear, the Hevelian scheme had been overshadowed. Although he does not say so directly, his concern with "memory" may have suggested Riccioli in more than one fashion. Claiming the lunar surface as a memorial to its discoverers (interpreted broadly) seems apt for an observer who aimed at taking his own place in this pantheon. Riccioli's scheme must have been seductive to all who sought a place in astronomical history. This, in the end, was its true success.

13

A Lunar Legacy

NAMES AND THE PLANETS

SPHERE OF INFLUENCE

*W*hether at a conscious or unconscious (cultural) level, astronomers of the eighteenth and nineteenth centuries used the seventeenth-century Moon as a model for their own nomenclatural forays into the solar system. Improvements in telescope design and technology, especially by William Herschel in the 1780s, brought a number of older worlds, such as Mars, Mercury, and Venus, into new and closer perspective. Surface features became visible (or at least it appeared so). At the same time that the solar system seemed to be shrinking, it was also expanded unexpectedly with the discovery of *new* planets, first Uranus and then Neptune. This was recognized as a major event in human history: not for millennia—since before the Greeks, the Babylonians, and even the ancient Egyptians—had a new planet been added to the canon of the "wandering stars." And not one, but two such additions had taken place within a mere sixty-five years, the life span of a single individual. The moderns had thus more than proven their superiority with regard to the heavens. The responsibilities of naming were the spoils of discovery.

URANUS AND NEPTUNE

The discovery of Uranus by William Herschel in 1781 was an event that rocked the astronomical world. Suddenly, it seemed, the solar system was a different place. Herschel entered history with a magnificence and an abruptness barely seen since Galileo. He had proven beyond any doubt that the telescope remained a crucial implement of discovery, not just observation and measurement. Indeed, the connection may not have been lost on him for he followed directly in Galileo's footsteps in another manner as well.

As a result of his discovery, King George III granted Herschel a considerable pension and a residence at Windsor, allowing him to devote his future energies entirely to astronomy. "I cannot but wish," wrote Herschel soon afterward, "to take this opportunity of expressing my gratitude by giving the name . . . to a star, which (with respect to us) first began to shine under His Auspicious Reign."[1] With Enlightenment restraint, the author of these words echoed those of his Tuscan predecessor: "But now, Most Serene Prince . . . behold, therefore, four stars reserved for your illustrious name." Herschel titled his new planet "Georgium Sidus," following Galileo's "Medician stars." Yet Galileo is not the only one to be evoked here. Michael Florent Van Langren, the first to publish a full-scale map of the Moon, had placed more than one of his patrons—both economic and spiritual—on the lunar surface. As noted earlier, the name of Philip IV, then king of Spain, appears five times on this map.

None of these three nomenclatural choices survived. Van Langren's Catholic Moon was occluded by Hevelius's classical geography and ultimately by Riccioli's astronomical community. Galileo's "Medician stars" remained in place for most of the seventeenth century but thereafter gave way to their rivals, coined by Simon Mayer, who discovered them only months after Galileo: Io, Callisto, Ganymede, and Europa, the most well known of Jove's mortal "possessions." On the Continent, "Georgium Sidus" was known for a while as "Herschel" (particularly after reports of George III's insanity became common) before being changed to "Uranus," a title finally standardized by Johann Bode at the turn of the nineteenth century. Thus, attempts to insert a modern into the ancient group of "wandering stars" all failed; in a new era of growing national tensions, astronomers eventually saw the benefit of holding to the tradition of Roman deities. Uranus therefore "belonged" to no one state or person.

Before this final determination was made, however, a similarly short-lived coinage was advanced for the next planet discovered, Neptune. The story of this discovery, with all its varied characters and details, has been told many times.[2] Rivalries among British, French, and German astronomers and unfortunate official blundering at the Greenwich Observatory lie close to the heart of it. In 1842 the Göttingen Academy of Sciences offered a substantial prize to anyone who could determine the reason for the growing perturbations in Uranus's orbit (caused by Neptune, of course). This problem was solved independently by Urbain Jean Joseph Leverrier in France and John Couch Adams in England. Adams, a young Cambridge graduate without an official post, had the earlier claim (September 1845, a full year ahead of Leverrier). He was not taken seriously, however, by those in charge at Greenwich, who could have used his calculations to locate the planet. Leverrier, on the other hand, was an assistant to the great Laplace and was widely recognized as his probable successor. He was assigned the Uranus problem by the director of the Paris Observatory, François Arago. Upon completing his work in September 1846, he immediately wrote to the Berlin Observatory; within several days the new planet had been confirmed. Had Adams himself written to Göttingen, he would have been credited with the discovery. As it was, his unofficial status earned him a snubbing from his British colleagues, whereas the more legitimized Leverrier received all the support he required.

This libretto with its sad conclusion was replayed on the level of nomenclature. Leverrier, in the midst of French celebrations over his success, took the liberty of naming the new planet, and he first chose the name "Neptune." Very soon thereafter, however, he did an about face and sought to have the world named after himself, "Leverrier." He did this apparently out of chauvinism: Uranus was still being called "Herschel" on the Continent. It seemed only right that Britain should not be allowed the only entry into the canon of planetary deities. With rather embarrassing strategy, Leverrier claimed in the published version of his findings, *Recherches sur les mouvements d'Uranus,* that he would henceforth never use the name "Uranus" to refer to the planet "Herschel."[3]

THE CASE OF MARS:
A NINETEENTH-CENTURY BATTLEGROUND

After the Moon, Mars is the most easily observed world from the Earth. As early as the 1630s, markings on its surface were described by Francisco Fontana, an early maker of lunar images. Hevelius mentioned such markings, too. A little more than a decade after *Selenographia* appeared, Huygens published the first drawing of the Martian face, actually a sketch, showing a central dark region later called the "hourglass sea." This sketch looks remarkably like Kepler's own drawing of the Moon in his *Dioptrice* or Thomas Harriot's first lunar image and is a reminder that all three men were mathematicians. Huygens's image, however, suggested to most seventeenth-century observers that Mars was indeed similar to the Moon, and thus, to the Earth as well. It had seas (dark areas) and lands, and therefore probably life. Such beliefs only deepened after 1666 when Cassini identified whitish areas around the polar regions, which suggested ice caps, and when William Herschel noted that Mars's equator was tilted twenty-five degrees, nearly the same as the Earth's, implying that the planet also experienced seasons.

It wasn't until the mid-nineteenth century, however, that the first naming schemes were applied to Mars, after the publication of an excellent map drawn in the 1830s by Johann Mädler and Wilhelm Beer, the two scientists who had also produced a standard-setting lunar atlas. In 1850 an Italian astronomer, Father Angelo Secchi, took Beer and Mädler's map, outlined major continents and seas, and added the names of famous explorers, such as Marco Polo, Cook, Columbus, and Cabot. Secchi created what, in his own day, was the equivalent of Gilbert's lunar nomenclature based on a naked-eye drawing. Both men perceived a New World in the skies. Whereas Gilbert employed the old Ptolemaic system of descriptive titles (Sinus Magnus, Insula Longam, etc.) that were still used in his time, Secchi adopted the approach of the high colonial period, which included enshrining individual explorers and discoverers.

Secchi's scheme did not last very long and was not widely followed. A decade and a half later, in 1867, it was replaced by that of the British astronomer Richard Proctor, who produced a map showing oceans, continents, and seas bearing dozens of new names. This nomenclature seems based on that of Riccioli since it dubbed features after astronomers who had contributed to the observation of Mars. A slight problem was noted, however: out of more than forty names, twenty-two were British and several were those of living scientists. Some names were used more than once: for example, five separate features received the title of "Dawes" (after W. R. Dawes, whose map Proctor had used to produce his own). Proctor, however, apparently became aware of his mistake. In a later version of his map, dated 1888, he added new titles that were modeled more directly on the Riccioli/Grimaldi nomenclature: Phaethontis (Storm Land), Windy Land (in English), and Niliacus Lacus (Lake of Gloom), among others. Such titles, with their imitative meteorology, had no direct ties to the cultural associations for Mars, which were far less prominent than those for the Moon, whose Sea of Rains and Ocean of Storms were linked to contemporary beliefs about lunar influences. Proctor's scheme was seriously challenged even before the ink on it had dried.

By the late 1870s, the next major proposal was on the table, this time offered by another Italian astronomer, G. V. Schiaparelli. Schiaparelli, in fact, presents a stunning correlative to Hevelius. He, too, was an observer of the first rank, was an excellent draftsman, and spent much of his career producing *Selenographia*-like treatises of the planets Mars, Mercury, and Venus. His map of Mars was certainly superior to anything that had come before and indicated that the Martian surface was in all likelihood very different from that of the Moon. Finding Proctor's map and nomenclature too imprecise, he decided upon another scheme with names that would be easy to remember and "whose sound [would awaken] such pleasant memories" in its users.[4] The scheme was that of Hevelius: it focused on names from classical geography centered on the ancient Mediterranean world. As analyzed by Jürgen Blunck, the total nomenclatural span that Schiaparelli had in mind is more elaborate than that which Hevelius conceived:

> [T]he dividing line on Mars reflects the day's journey of Helios from the East to the West. Having slept in a beauteous mere (Solis Lacus) on Oceanus, Helios starts his journey in the morning by fabulous east Indian islands Aurea Chersonesus (Malacca), Agathodaemon (Sumatra), Chryse (Thailand) ... passes along the south coast of India (Margaritifer sinus) . . . the Arabian Sea (Fastigium Aryn) . . . the North Africa coast (Aeria, Syrtis maior, Libya . . .), to the far West, the region between Aethiopis, Atlantis, and Herculis columnae.

In the same way the latinized names for the Old World stretching to the furthest points of Greek exploration encompass the cultural spheres of the Bible in the East (Tharsis, Ophir, Eden . . .), of the Egyptian religion around Aeria (Isidis regio . . . Thoth), and of the underworld mythology toward sunset . . . with Trivium Charontis at its center. From here canals with underworld names stretch all over the large continental block west of Elysium.

Moreover, various designations along the dividing line on Mars are reminiscent of the wanderings of Odysseus . . . : first sailing from Troy/Hellespontus; then with the Lotus-eaters near the Syrtis minor; then in the land of the Cyclopes (Cyclops); then in the island of Aeolus (Aeolis); then with the Laestrygonians (Laestrygon) . . . ; then with the Sirens (Mare Sirenum) . . . ; and finally returning to Ithaca/Hellas.[5]

Schiaparelli, following Hevelius, believed he saw in the features of Mars the classical world of heroes and myth. This was the world that would awaken "such pleasant memories," a world no longer populated by the ancients and the challenges they presented to a modern, but instead a world of schoolboy charm, touched with the nostalgic elegance of an elite upbringing.

Nonetheless, a debate flared up all over Europe about which system was more appropriate for the future of astronomy, Schiaparelli's or Proctor's. Even the British were divided: some favored Proctor's scheme because it was already widely used; some argued for Schiaparelli; others recommended using letter designations only; still others promoted "recourse to the mythological dictionary, and [use of] the names of male characters [so that] we shall not be intruding on the property of the minor planets."[6]

By the last decade of the nineteenth century, the tide had clearly turned in favor of Schiaparelli's system. This was secured when the system was adopted by two influential astronomers who would largely dominate Martian observation—Percival Lowell (of "Martian canals" fame) in the United States and Eugène Michael Antoniadi in France. Antoniadi spoke of "Schiaparelli's charming aerographical nomenclature." Lowell was even more effusive in his praise, describing it as being "at once [an] appropriate and beautiful scheme, in which Clio does ancillary duty to Urania."[7] Whatever Lowell's idea of "history" (Clio) might have been, invoking Clio and Urania was appropriate because it suggested the premodern qualities of Schiaparelli's nomenclature. Although Proctor had included living astronomers in his scheme and thereby tried to trump posterity, this was not why his system failed.

Why, then, was the astronomical community in 1890 so ready to accept a scheme for Mars that it had soundly rejected a century earlier for the Moon? What made Schiaparelli's naming system attractive and successful, even "charming," whereas that of Hevelius had seemed "bizarre" and overly complicated? The answers are themselves complex and have everything to do with nonmythological history, but they are not so difficult as to prevent some fruitful generalizations. In Hevelius's day and well into the eighteenth century, the ancients represented not only a body of knowledge to be admired and learned, but also, among scientists free of the Church's commands, to be surpassed and even rejected. Galileo, Kepler, and Newton had all done as much and helped set the standard; indeed, Galileo had been adamant about overturning Aristotelian views of the heavens, and Kepler had dealt the final blow to Ptolemy. Yet debates continued to rage, especially in France and England, about whether the moderns were as yet worthy to stand beside the ancients. Jonathan Swift's satiric pamphlet, *Battle of the Books* (1704), ridiculed this controversy near the peak of its heat but hardly cooled it.

Latin was declining as the universal language of knowledge, and scientists sought public recognition in the vernacular for their hopes of patronage. The astronomical community, like much of science at the time, was scattered among the members of a few professional organizations (e.g., the Royal Society) and a large number of self-taught amateurs (often very skilled ones). This key role of amateur astronomers continued well into the next century, as exemplified by William Herschel, who was a musician by background. By the mid-nineteenth century, however, astronomy had become a solidified community of trained professionals, with an institutionalized sense of its own. The new equipment of astronomers, their achievements, and their status as "scientists" (a word coined in the 1830s by William Whewell) placed them so far beyond the reach of previous centuries and so deeply into the modern that the ancients had become beautiful relics, full of moral and spiritual example as the bearers of a "golden" history and literature but largely bereft of true scientific value. There were exceptions, of course; Euclid and Archimedes come to mind. For the most part, however, the authorities who had ruled the heavens up to and even beyond the Renaissance—

Aristotle, Pliny, Ptolemy, Capella, Macrobius, and their commentators—had been reduced to "philosophers" or "literary" authors. Ironic as it may seem, Schiaparelli's nomenclature for Mars appealed to both of these qualities attached to the ancients. The "charm" and "beauty" of his scheme resided in a sense of sacred antiquarianism, as Lowell expressed it, mixed with a less evident quality of chosenness. By baptizing Mars as an evocation of the classical world, Schiaparelli was memorializing the astronomical community's own self-image. In Hevelius's day, this image bore a different meaning altogether: it implied that the ancients were still superior to the moderns. Two hundred years later, these same ancients had been put on the shelf, reduced to symbols of high culture.

MERCURY: HEAT AND LIGHT

Schiaparelli produced the first map of Mercury. After assembling as many as 150 drawings of the planet during the 1880s, he published a final map showing surface markings that he said were reminiscent of the lunar maria.[8] In fact, the map showed features that resembled streaks more than spots, but Schiaparelli chose to stress the lunar analogy without adding the specificity of names. He also claimed a captured rotation for Mercury, again like the Moon. Johann Schröter believed he had observed towering peaks on Mercury that rose to heights above sixty thousand feet; he called them "the Cordilleras." Other than this, however, the planet had attracted no real nomenclature. This task was left to Antoniadi, who produced his own map in 1933, very much along the lines of his predecessor's; it remained the standard version until 1965.

Not surprisingly, Antoniadi's naming scheme was given in Latin and found its inspiration in classical mythology. Taking his cue from Schiaparelli—but more historically, from Hevelius—Antoniadi sought a general theme. Because of Mercury's proximity to the Sun and his conviction that one face of the planet was forever turned toward the solar flame, he conceived the existence of vast desert tracts—Solitudo Hermes Trismegisti (for the Egyptian god of gnostic powers); Solitudo Promethei (for the Titan banished to the dry peaks of the Caucasus); Solitudo Atlantis (for the mythic island, said to be favored by the Sun); and Solitudo Persephones (for the maiden whose six-month annual stay in the underworld caused the Earth to be barren during that time). To areas he identified as possible dried up seas, Antoniadi gave titles that spoke of heat or brightness: Apollonia, Pieria, Phaethontias, Caduceata, and Heliocaminus, among others.

Unlike Mars and the Moon, Mercury was not overtly forced into the mold of a specific terrestrial geography. Yet it could not escape the designations of "territory" and "place" that had been put upon these worlds—lands and seas above all. The pattern begun with the lunar surface was unbreakable; the analogies were too ready at hand. If the Moon had become another world in its own right as a result of being perceived as another Earth, then Mars, Mercury, and the other planets would gain their cartographic reality through perceived similarities to the Moon. For modern astronomers, it was much safer to call upon this comparison than upon the one for which it served as an alibi. Saying that Mercury was like the Earth was patently absurd because of its much smaller size, its presumably captured rotation, and its terrifying closeness to the Sun. Saying that Mercury resembled the Moon, however, not only avoided this difficulty but it also suggested that the solar system was a cozier realm than previously thought, a domain composed of spheres that played variations on a few crucial themes.

CONCLUSION

The cursory survey presented in this chapter suggests that the discovery of the lunar surface in the seventeenth century established a schema for planetary observation that lasted three hundred years. Between the reconnaissance lunar drawings of Galileo in 1610 and those of Mercury by Schiaparelli in 1889, between the finished maps of Hevelius and those of Antoniadi, there is more than sympathy; there is visual echo. The great distance of time, the vast improvements in telescope design, the advent of modern observatories, and even the use of photography—none of these factors separating the centuries in which these images were drawn were sufficient, either by themselves or in combination, to diminish the use of the Moon as the model for perception. The Moon, however, is really the Earth under mask.[9]

As its nearest and most observable neighbor, the Moon became the Earth's alter ego in the Western imagination, and in turn, exerted its influence on the rest of the solar system. The lunar face proved itself the sometime standard for even the Sun, as shown by an engraving in Kircher's *Iter exstaticum,* which depicts the solar surface full of round, flaming craters from which black clouds of smoke (sunspots) emerge.[10] The Moon did not merely set a pattern; it created predispositions that awoke each time "discovery of place" was reenacted in the heavens.

The power of naming is the power to create new geographies of vision, place, and knowledge. Making claims on distant objects, however, especially on one as ancient and culturally burdened as the Moon, can not be done too parochially without generating rejection (as the cases of Van Langren, Hevelius, and Proctor make plain). The Moon was reborn with Galileo's discovery of its surface, yet it inherited imageries that had existed for thousands of years. It was Kepler who reminded Galileo of his debt to Plutarch, fourteen centuries his elder. Had he written further on the matter, Kepler might have observed that Galileo's defeat of the Aristotelian position echoed Plutarch's intention in proposing an Earth-like Moon, and that this, in turn, drew on debates and images that had existed for nearly a thousand years. The conditions generating the impulse within each age to perceive a terrestrial Moon were, of course, wholly different, as was the forum of expression. But what may be the most primordial image of the lunar surface, indeed of all planetary surfaces, was not to be denied its destiny of eventual proof. Such proof came at the hands of astronomical science in its first blush of modern observational discovery. It was not completely overturned until human beings actually set foot on the Moon. Only then, when the lunar surface became true landscape and entered into living rooms all over the Earth, was it clear that this world was a wholly alien, distant place, unlike anything found in the terrestrial domain.

14

The Lunar Cycle

nalogy is at once the cruelest and the most magnanimous of intellectual masters. Its demands can never be satisfied, yet it is infinitely forgiving and generous toward all ideas and perceptions that appear at its door. Moreover, these demands are continually rejuvenated as a result of the very gaps and insufficiencies that gave them life. In the case of the Moon, the success of the earthly analogy may have been predicated on its magnificent utility to many ages in many ways. Such utility extended far beyond the bounds of making the distant and unfamiliar a stage for the intimate and perceptible. If it served as an "isle of the blest" for the Pythagoreans or a realm of "many mountains, rivers, cities" in the Orphic rhapsodies, this was perhaps because of a need to propose a visual reference for the religious effects of loss associated with death. For Plutarch, who was prone to perceptive moralizations in an age of lavish display, an Earth-like Moon was a call to humility for those who believed that all beauty and landscape ended in the sublunary realm. Philosophers who perceived the Moon as a mirror, rag of flame, or crystal ball were as provincial as the emperor who never left his palace (with or without clothes).

In medieval European astronomy, the Earth-Moon idea did not take hold until very late; the reasons for this are not simple. It is perhaps true that this type of conception was not well suited to Christian belief systems, which viewed the heavens as useful for their revelations of time, divine influence and order, and the like. The computistic-mathematical heavens, which dominated one side of medieval astronomy, were another entry into the book of comparisons—as Plato advised, all geometry returns to the earthly sphere at the center. But this was euphemism; the lunar surface held far more interest for Arab astronomers and philosophers such as al-Haytham (influenced as they were by Greek thought) than it ever did for Europeans prior to the twelfth and thirteenth centuries.

Instead the Moon first achieved its own detailed visual reality through the brilliance of art. It was the matching of this brilliance with that of the new astronomy of discovery that led to the Moon's terrestrial assumption. It was in this union most of all where allegory could become a basis for realism. Paintings of Selene (Cynthia) or of the "man in the Moon" held the lunar disk on a symbolic string, but they also established it as an object of *drawing,* not just of mathematical illustration. This new status promised that the Moon would eventually be given its place in the repertoire of objects available to the observing, recording eye. Such did not happen easily, however. Although observing and documenting nature had been practiced for centuries, ever since the early phases of Gothic naturalism, it still took a Van Eyck to extend it beyond the parochial (if enthralling) limits of the organic into the full realm of physical reality. Van Eyck included the Moon among the objects of the Earth. No clearer evidence for this exists than the fact that he painted his clearest portrait of the lunar face, complete with all its maculae, in the daytime, in the midst of a museum of terrestrial visibility, just beneath the cross on which one of the thieves suffers in crucifixion: there it hangs, casually, part of the moment, without symbolic fastenings.

Galileo's images of the Moon carry forward Van Eyck's materialism with a new and highly specific purpose. Yet, as shown by William Gilbert, even by this time the lunar surface had been made the subject of cartography, not just picturing. Epistemologically, it was this tiny yet bold stroke that marked the great break with the past: Gilbert appears to have been the first to codify perception of the Moon as "territory" and "place" into the form of an actual map. The map, of course, implied travel, and the vessel by which science finally journeyed to the Moon and took control over it was the telescope. It was through this instrument and its complement the microscope that the representational imagination gained a central place in science during the early seventeenth century.

The telescope was the crucial impetus for science and art to join forces in the modern world with regard to the heavens. Both telescope and microscope, as extensions of the eye, demanded a reporting of "sight." They did not merely strengthen human perception, however; they reconstituted and transformed it as well. They did this not only by deepening the eye's horizons, by creating a host of new visibilities whereby the world and the skies were reformed into a universe of details; but they also did it, with regard to the Moon especially, by making ancient concepts visible or apocryphal. The idea of the Moon as another Earth, complete with mountains, seas, and islands, was a vision that had to be grasped, literalized as perception, and entered into the canon of images-as-facts. More than this, it became itself a means by which an evolving present could claim superiority: Galileo's little book, *Sidereus nuncius,* was a formidable act of conquering in its own right, for it proved that a living generation of thinkers could achieve unprecedented advances and thus enter the geography of fame previously reserved for the ancients. It was now evident that neither Greece nor Rome nor the Church Fathers had known as much about the physical universe as did contemporary thinkers. With the world and the heavens drawn closer by means of documented perceptions, the ancients receded. "It is a striking fact," said a writer of the late sixteenth century, "that our classical authors had no knowledge of all this America, which we call New Lands."[1] Such was even truer of the Moon.

A QUESTION OF ORIGINS

This brings us to certain epistemological concerns. It has been said many times by many authors that seventeenth-century observers responded to the microscopic world in a very different way than they did to the cosmos revealed by the telescope. The "two infinities," as Pascal called them, were not mirror images of each other and did not circle back to meet in any obvious manner. Very few observers who entered the depths of one of these worlds ever ventured far into the other before turning back. The one, it is said, evoked a kind of scientific horror, the other, a pleasure sometimes bordering on rapture. As one famous historian of the last century commented:

Nothing is more curious than to observe the utterly contrary impressions that the two revolutions enacted upon their originators. Galileo, standing before the infinity of the heavens, where all appeared harmoniously and magnificently calculated, felt joy even beyond surprise; and he announced everything to Europe in a style all the more playful. Swammerdam, before the infinity of the microscopic world, seems to have been seized with terror. He retreated before the abyss of nature in combat, devouring itself . . . fearing that all his ideas, his beliefs would be rendered fragile.[2]

The author, Jules Michelet, seems bemused at the contrast, yet more recent studies have taken the matter seriously. A recent work, for example, concludes in much the same fashion but with a different tone, stressing that the century of the Scientific Revolution witnessed a "sense of dislocation induced by the discovery of the microworld," even though it was softened with an "enchantment by images."[3]

Michelet seems correct with respect to Galileo's picturesque language. His description of dawn rising over the lunar peaks and plains reveals no fearful "dislocation" of any sort. On the contrary, his trend-setting narrative and the pictures he produced of the lunar face to illustrate it enthroned the familiar world of mountains, seas, and plains as the model of what might be termed "the Universal Planet." However displaced it became from the center of the solar system because of Copernicus, the Earth gained a new centrality in the seventeenth century that perhaps it has never lost. It was one of the achievements of the Scientific Revolution to "prove" and therefore sanctify this image of the Moon as Earth—an image that is among the very oldest and most cherished of the ancients.

In a justly famous book on the formation of the "scientific spirit" in the West, Gaston Bachelard opens with a striking statement: "To make representation geometric, in other words to make drawings of phenomena and to place in an ordered series the decisive events of an experience—such is the first task in which the scientific spirit affirms itself."[4] This type of image, the author says, hovering between "the concrete and the abstract" and seeking to reconcile "laws and facts," was of enormous power. It had power simply because of its "naïve realism," which would soon prove insufficient and thus demand to be superseded. Bachelard may have been thinking of Galileo when he wrote this, but he might just as easily have been speaking about William Gilbert, Johannes Hevelius, Michael Florent Van Langren, or Giambattista Riccioli. Each of these men sought a primitive realism for the Moon in the form of the map, the image that most decisively fills Bachelard's description.

For Galileo it was exactly those pictures of the "decisive events" of his own observational experience that made him a well-known scientist in Europe. He is stereotypically known today as a crucial originator of modern science for inserting mathematics into the heart of experiment, into physics. Before he did this, however, he was the framer of a new realism; this was his most crucial contribution to astronomy at the beginning of the new century. Van Eyck, Leonardo, Gilbert, Galileo, and other makers of lunar pictures showed that images, no less than mathematics or the "experimental philosophy," lay at the core of the Scientific Revolution. The deeper content of these images, culminating in Galileo's own pictures, was this: they were radioactive with *theory*. In their very forms, they carried a specific hypothesis about the nature of the lunar substance, the same hypothesis that was everywhere extolled in text, whether by literary or astronomical authors. Thus the lunar imagery of Galileo, Hevelius, and Riccioli, as well as of those that came after, was not an aid to science or a helpful addition. It *was* science, ripe with the mixture of fact, form, experience, and interpretation that Bachelard would tell us is the more complete expression of the "scientific spirit."

OF NAMES AND THE MAN

Van Langren, Hevelius, and Riccioli all proposed an Earth-like Moon by virtue of their maps, but it was not quite the same Earth that each of them used as a model. The nomenclatural systems of these men were all published in a very brief period of only six years between 1645 and 1651, during which the entire balance of power in Europe shifted away from Spain and papal authority to northern Europe in defiance of Rome.

Van Langren's scheme stained the Moon with the existing world of politics and patronage. Its own favoritism, which belittled science in more than one way, could not hope to succeed. Hevelius, on the other hand, made of the Moon a billboard for an outmoded classicism; it produced another geography vacant of science. It was Riccioli who finally took fuller possession of the lunar surface for astronomy. Yet this "science" was not a wholly stable thing; it brought together ancient, medieval, heliocentric, mythological, and religious ingredients. Riccioli, like Van Langren and Hevelius, included allegiance to his ultimate patron (Rome); his "treatment" of those such as Copernicus and Kepler makes this evident. Yet he did not shrink from something far broader and more appealing to his time. Rewriting the Moon as a text of its own ecumenical past was a great act, and we are fortunate to have it preserved today. It reveals directly that any "science," be it physics or astronomy, is a discourse of history as much as of theory, fact, or hypothesis.

The eventual success of the Riccioli/Grimaldi nomenclature meant that science would inherit a mid-seventeenth-century Moon. This was finally made doctrine in 1935, when the International Astronomical Union released its report standardizing the names of nearly six hundred lunar formations in accordance with this scheme.[5] Further updated and greatly expanded during the 1960s, the official nomenclature continued to build on the Riccioli system but narrowed it to the names of deceased scientists only.[6] At the same time, the use of Latin was strictly adhered to, with many new terms being added to designate new types of observed features (e.g., *rupes* for "scarp"). These were also applied to the other planets and their moons, effectively casting the whole of the solar system into the lunar shadow once again. Since the 1980s, a great many non-Western names—of artists, writers, musicians, gods and goddesses, towns and cities (ancient and modern)—have been added to the pantheon of titles. Not, however, for the Moon.

Riccioli's scheme can not be deprived of its magnificent topography of association. However amended or expanded upon today, its names reveal the myriad voices and the complex range of historical beliefs, superstitions, theories, and ambitions that have gone into the making of the Moon as a cultural object of the imagination. The oldest earthly geography of the skies remains a historical document of the age that "discovered" it.

Though studied and investigated today by people throughout the world, the Moon will always speak of Europe first and of the politics of European science in the mid-seventeenth century. This, however, should be cause, not for dismay or simple denouncement, but for historical perception. The Moon of today is the result of troubled adventures of the eye, hand, and mind that went far beyond the limits of the terrestrial realm, only to be discovered again in rarefied form.

A BEGINNING WITHOUT END

We might end where we began, with the rest of the solar system. Mercury, Venus, the moons of Jupiter, Saturn, and Uranus: all these have now been given their naming schemes. These schemes at first seem diverse: famous deceased artists, musicians, painters, and authors from all cultures for the craters on Mercury; goddesses and famous women for Venus; astronomers who discovered Jovian satellites for Ganymede; people and places from Homer's *Odyssey* for Tethys; and so on. Such is our library of names on other worlds. It seems wholly contemporary in its multicultural reach. Yet a few moment's pause over the systems applied to the Moon three centuries ago will show that there is nothing here that is new under the Sun.

Will the Earth one day relinquish its hold on the Moon as its model for other planets, as its own alter ego? What uses would such a lunar world have, culturally speaking, were it allowed to die? Beyond nomenclature and simple geography, the Moon has been irretrievably linked to ideas of the distant voyage. It has been colonized over the centuries by assumptions of terrestrial realities far beyond those discussed in this book: life, cities, castles, wars, volcanic activity, an atmosphere, swamps, marshes, gaseous extrusions, and more. If there were many Moons in the past, many different versions of the lunar orb and its meaning, this is because there were just as many Earths that this orb helped bring to light.

Only a few precious years before the first man set literal (not literary) foot upon the lunar surface, many of the beliefs about the Moon from long ago were still in evidence among astronomers in one form or another:

> Seasonal changes [on the lunar surface] there certainly are: some markings darken, others become paler, expand or contract, with a variation of hue, in the course of the lunar day; nor are these changes symmetrical as between evening and morning. The temperatures of the topsoil are in step with the phase. Most of the seasonal changes, on the other hand, lag behind the Sun two or three days, keeping pace with the subsoil temperatures. This shows that they do not depend on superficial alterations but have their seat some way below the ground. Such indeed would be the behaviour of vegetation sending long taproots down into the gas marsh.[7]

We can only read the title of this work, *Strange World of the Moon,* with ironic finality and understanding.

When will humankind abandon the notion of an Earth-like Moon? The answer is simple. If today we look for water on the lunar surface (and what more promising substance of life-as-we-know-it than this?), tomorrow we may search for some other optimistic material—perhaps methane, carbon dioxide, or polycyclic hydrocarbons. These assumptions have their own historical meaning. They reflect back sherds of meaning about how we view ourselves today at the center of the universe. If this universe is not the same as it was even a few decades ago; if it has become more grandiose, violent, and explosive; we can at least be comforted by the truth that the most ancient dream of the lunar face will not pass away. It is a dream from which we will never wake. Our first and nearest vision in the heavens shall always be our closest companion.

Notes

CHAPTER 1. THE FIRST MODERN PLANET

1. A wealth of information on the IAU, its membership, and its history can be found in Blaauw, *History of the* IAU.

2. Skelton, *Maps,* p. 3.

CHAPTER 2. HOW THE MOON BEGAN

1. Visual tests have been devised to measure the ability of the unaided eye to distinguish features on the lunar surface (see, for example, Ashbrook, *Astronomical Scrapbook,* p. 233). Comparisons suggest that, at the time of maximum visibility (early morning twilight, when night glare is absent), nearly as much of the surface can be discerned as is apparent in the first telescopic views drawn by Galileo.

2. Murdoch, *Album of Science,* p. 244.

3. The scholarly problems surrounding the lack of original source material for the pre-Socratic philosophers have been discussed in Havelock, "Linguistic Task of the Presocratics," and in Osborne, *Rethinking Early Greek Philosophy.*

4. For my own survey, I have drawn on a number of these sources, as represented by twentieth-century collections and translations. These include Diels, *Die Fragmente;* Guthrie, *History of Greek Philosophy;* Heath, *Greek Astronomy;* Wheelwright, *Presocratics;* and editions of classical authors, mainly published in the Loeb Classical Library series (Aristotle, Diogenes Laertius, Hesiod, Lucian, Plato, Plutarch, etc.). Because of its wealth of information and references, I have also consulted Wilkins, *Discovery of a World.* This work played a very important part in the popularization of lunar discoveries during the seventeenth century.

5. Evelyn-White, trans., *Homeric Hymns and Homerica,* pp. 371, 459, 461.

6. Diels, *Die Fragmente*, p. 33.

7. See, for example, Kerenyi, *Die Mythologie*, especially pp. 186–201.

8. Guthrie, *Orpheus and Greek Religion*, p. 73.

9. Ibid., p. 69.

10. Ibid., p. 138. For the original Greek, see Kern, *Orphicorum fragmenta*, p. 26 (fragment 91).

11. Guthrie, *History of Greek Philosophy*, vol. 1, p. x.

12. Quoted in Wheelwright, *Presocratics*, p. 51.

13. Wheelwright, *Presocratics*, p. 57. For a slightly different translation, see Kirk et al., *Presocratic Philosophers*, p. 135.

14. Freeman, *Presocratic Philosophers*, pp. 69–70, and Heath, *Greek Astronomy*, pp. 10–11.

15. Wheelwright, *Presocratics*, p. 83. The quote is translated from the third-century A.D. writer Diogenes Laertius, whose work *Lives of the Eminent Philosophers*, despite its many flaws and general gullibility, has long been a valuable source of information on early Greek thought.

16. Dreyer, *History of Astronomy*, p. 21. See also Kirk et al., *Presocratic Philosophers*, pp. 200–201.

17. Wheelwright, *Presocratics*, p. 100. See also Guthrie, *History of Greek Philosophy*, vol. 2, p. 66. Plutarch, in *The Face in the Moon*, quotes this line of Parmenides' as "fixing her glance forever on the sun" (see 929e, p. 101).

18. As quoted in Plutarch, *The Face in the Moon*, 929d, p. 103. See also Guthrie, *History of Greek Philosophy*, vol. 2, pp. 197–198.

19. Dicks, *Early Greek Astronomy*, p. 54.

20. Aratus, *Phænomena*, p. 267.

21. Plato, *Timaeus*, 38 b ff, 39, in *Collected Dialogues*, pp. 1167–1168.

22. Plato, *Laws*, X, 887 d, 899 b, in *Collected Dialogues*, pp. 1442, 1454.

23. Ibid., X, 887 d, p. 1442.

24. Ibid., XII, 967 a,c, pp. 1511–1512.

25. Plato, *Timaeus*, 42d, p. 1171.

26. Ibid., 33b,c. For this passage, I have drawn from the versions presented in *Collected Dialogues*, p. 1164, and in Heath, *Greek Astronomy*, pp. 49–50.

27. Freeman, *Presocratic Philosophers*, p. 206.

28. Guthrie, *History of Greek Philosophy*, vol. 2, p. 308; Dreyer, *History of Astronomy*, p. 189.

29. Freeman, *Presocratic Philosophers*, p. 268.

30. Heath, *Greek Astronomy*, pp. xxxv–xxxvi. See also Wheelwright, *Presocratics*, p. 164.

31. Plato, *Apologia*, in *Collected Dialogues,* pp. 12–13.

32. Freeman, *Presocratic Philosophers,* pp. 279–281; Guthrie, *History of Greek Philosophy,* vol. 2, pp. 370–372; Dicks, *Early Greek Astronomy,* pp. 77–78.

33. Dicks, *Early Greek Astronomy,* pp. 65–67; Dreyer, *History of Astronomy,* pp. 40–42. See also Lloyd, *Early Greek Science,* pp. 27–28.

34. Guthrie, *History of Greek Philosophy,* vol. 1, p. 285.

35. Ibid., p. 285, n. 3.

36. See, for example, Clagget, *Greek Science in Antiquity;* and Neugebauer, *Ancient Mathematical Astronomy,* vol. 1, pp. 3–5.

37. Good summaries, complete with diagrams, of this and other geometric systems proposed for the planets can be found in Clagget, *Greek Science in Antiquity,* pp. 86–98; or in North, *Astronomy and Cosmology,* pp. 66–119.

38. Neugebauer, *Ancient Mathematical Astronomy,* vol. 1, p. 5.

39. North, *Astronomy and Cosmology,* p. 68.

40. Dilke, *Greek and Roman Maps,* p. 22. According to Dilke and other authors, the first Greek map of the known world was drawn by Anaximander in the late sixth century B.C.

41. O'Neil, *Early Astronomy,* pp. 52–82; and Neugebauer, *Exact Sciences in Antiquity.*

42. Murdoch, *Album of Science,* pp. 73, 244–245.

43. Aristotle, *De caelo,* p. 179.

44. Aristotle, *Parts of Animals,* p. 451.

45. Aristotle, *Meteorologica,* p. 19.

46. Heath, *Greek Astronomy,* pp. 123–124.

47. Ibid., p. 124.

48. This term is from Winkler and Van Helden, "Representing the Heavens."

CHAPTER 3. EPIC JOURNEYS AND FLIGHTS OF FANCY

1. Romm, *Edges of the Earth.* This work provides an excellent overview of the development of geography, as idea and practice, in classical antiquity.

2. Ibid., p. 112.

3. Dowden, trans., *Alexander Romance,* pp. 719–720.

4. This point is made by Romm, *Edges of the Earth,* pp. 131–132.

5. Dicks, *Early Greek Astronomy,* pp. 65–68. Apparently the Pythagoreans used both *antichthones* ("counter-Earth") and another term, *aitheria gaia* ("heavenly Earth") for the Moon (Dicks, p. 67). These two usages were distinguished in some manner, with the latter term reserved for contexts in which notions of "purification" after death were important.

6. Stephens and Winkler, *Ancient Greek Novels,* pp. 123–124. A note attached to this phrase indicates this was also attributed to the Celts (n. 45).

7. Ibid., p. 125.

8. Ibid., p. 124.

9. For this particular quotation, I have used the translation by Romm, *Edges of the Earth,* p. 208. For comparison, here is the version in Stephens and Winkler, *Ancient Greek Novels:* "Most incredible of all, that in their progress northwards they came close to the Moon, as to a very pure land, and there they saw what you would expect a person to see who had previously made up lies out of all proportion" (p. 126).

10. Translation by Cherniss and Helmbold in the Loeb Classical Library edition, *Plutarch's Moralia,* vol. 12, p. 35.

11. For example, Dreyer, *History of Astronomy,* says that the work contains "all that the most enlightened minds of antiquity could make out with regard to the constitution of the moon" (p. 189).

12. Lucian, *A True Story,* pp. 249–253.

13. Lucian, *Icaromenippus,* p. 275.

14. Translated by Lloyd in *Revolutions of Wisdom,* p. 329, n. 144. See also Ptolemy's *Almagest,* Book 1, chapter 1, p. 36.

15. Pliny the Elder, *Naturalis historia.*

CHAPTER 4. THE MOON AND MEDIEVAL SCIENCE

1. For recent summaries of this crucial episode in the history of Western science, see Lindberg, "Greek and Arabic Learning," and Montgomery, "Earthly Projections." A more extensive, if somewhat dated, standard source on this topic is O'Leary, *Greek Science.*

2. Eastwood, "Astronomy in Europe."

3. See the extensive commentary provided by Stahl et al., *Martianus.*

4. Stahl, *Roman Science,* p. 53.

5. Stahl et al., *Martianus,* vol. 2, p. 55.

6. For a different interpretation, see Stahl et al. (*Martianus,* p. 55, n. 113), who propose that these forms are associated with the different lunar phases.

7. These and other images, plus the myths behind them, are briefly described in Guiley, *Lunar Almanac.* Though not a scholarly work, this book contains a treasure trove of information on Moon mythology from around the world. Other useful works include Brueton, *The Moon,* and Wright et al., *To the Moon.*

8. A good reproduction of this image appears in Weitzmann, *Book Illumination,* p. 34.

9. The authoritative work on the Dura synagogue remains Goodenough, *Jewish Symbols.* Volumes 9–11 deal with the Dura images.

10. Goodenough, *Jewish Symbols,* vol. 9, pp. 110–116.

11. An example of such an image can be found in Bouché-Leclerc, *L'astrologie grecque,* p. 373; this image is repeated in Sarton, *Hellenistic Science,* p. 318.

12. Goodenough, *Jewish Symbols,* vol. 9, p. 117.

13. This image can be found in Weitzmann, *Book Illumination,* p. 18.

14. Reeves, *Painting the Heavens,* pp. 139–140.

15. Ibid., chapter 4.

16. These excerpts have been recently discussed by Eastwood in "Astronomy in Europe," "Plinian Astronomical Diagrams," and "Astronomies of Pliny."

17. McKitterick, "Text and Image."

18. Ibid., p. 300.

19. See, for example, Eastwood, "Astronomy in Europe."

20. An excellent study of this manuscript is offered in von Euw, "Die Künstlerische Gestaltung."

21. Examples of all of these are provided in von Euw, "Die Künstlerische Gestaltung."

22. Eastwood, "Plinian Astronomical Diagrams"; compare the drawings on pp. 150–162.

23. Ibid., p. 142.

24. Panofsky, *Renaissance and Renascences,* pp. 49–50.

25. A brief but excellent discussion of this manuscript, which includes a catalogue of its miniatures, can be found in Katzenstein and Savage-Smith, *Leiden Aratea.*

26. See Eastwood, "Leiden Planetary Configuration." The year A.D. 579 is believed to correspond to the date by which a copy of the Latin translation by Germanicus Caesar had been made. This copy was apparently used as the basis for the ninth-century rendition.

27. See Martin, *Histoire.* Page reproductions from this remarkable codex can be found in several publications, including Murdoch, *Album of Science,* p. 250; and Stott, *Celestial Charts,* pp. 40–41.

28. See Beckwith, *Early Medieval Art,* p. 48.

29. Quoted in Ganzenmüller, *Das Naturgefühl,* p. 79. This work remains one of the best references on the development of nature imagery from the end of antiquity until the twelfth century.

CHAPTER 5. THE LATER MIDDLE AGES

1. Summers, *Judgement of Sense,* p. 312.

2. Examples from the fourteenth century are mentioned in Classen, "Die Ältesten Mondkarte." The widespread nature of the "man in the Moon" image is evident in many sources, including works on astronomy and astrology; biblical commentary; and in the realm of art, stained glass windows (e.g., La Sainte Chapelle in Paris and "Joseph's Window" in the Erfurt Cathedral in Germany), painted church clocks, woodcuts and paintings by famous artists (e.g., Albrecht Dürer's *Apocalypse: The Signs of the Judgment,* used to illustrate literary works such as Sebastian Brandt's *Ship of Fools*). A broad variety of "man in the Moon" imagery can be found in a single work: Verdet, *The Sky;* see especially pp. 11, 14, 18, 21, 26, 45, 50, 56, 61, 74–75, 124–125, 127, 143, 152, 167, and 198.

3. See, for example, Haskins, *Renaissance,* and Benson and Constable, *Renaissance and Renewal.*

4. Thorndike, *Magic and Experimental Science,* vol. 2, p. 192.

5. See Ariew, "Galileo's Lunar Observations."

6. See Aristotle, *Generation of Animals,* Book 4, p. 10, and Book 3, p. 11.

7. Quoted in Ariew, "Galileo's Lunar Observations," p. 217.

8. Ibid., p. 219.

9. Ibid., p. 218.

10. See the summary discussion of his philosophy and the relevant bibliography in Sabra, "Ibn Al-Haytham."

11. Schoy, "Introduction."

12. A good example of this can be seen in the following passage, which also reveals the author's general view of the Moon as an optical entity:

> Roughness therefore hinders the reflection of light, but not its absorption; areas of unevenness absorb light most of all, because of how it penetrates into hollows and pores when it encounters a rough body, whereas smooth portions of the body reduce absorption. . . . Indeed, the smoother a body is the more its capacity for absorption differs from that of a rough body in the case of primary reflected light. (p. 11)

13. These ideas, including the noted experiment, can be found in *Paradiso,* Canto II. For a more extended treatment, see Dante's *Convivio,* Canto II, p. 14.

14. See the translation in Grant, *Sourcebook in Medieval Science,* p. 525.

15. Oresme, *Le Livre,* p. 459.

16. Quoted in Ariew, "Galileo's Lunar Observations," p. 221.

17. In addition to the recent study by Summers (*Judgement of Sense),* which

deals with the the later Middle Ages, other classics in this area include Mâle, *Gothic Image*; Goetz, "Die Entwicklung"; and Dvorák, "Idealismus und Naturalismus." Perhaps the best survey of the subject is White, "Natural Science."

18. Thorndike, *History of Magic,* vol. 2, p. 537.

19. Useful surveys of these developments can be found in Stock, "Science, Technology, and Economic Progress"; Whyte, *Medieval Technology;* Gimpel, *Medieval Machine*; and Long, *Science and Technology.*

20. de Bruyne, *L'Esthétique,* p. 93.

21. Quoted in Eco, *Art and Beauty,* p. 57.

22. Quoted in Ganzenmüller, *Das Naturgefühl,* p. 164. The passage describes the various stages of the soul brought close to the pure flame of divine love and contemplation.

CHAPTER 6.
THE FIRST DRAWINGS OF THE LUNAR SURFACE

1. Pächt, *Van Eyck,* p. 13.

2. See the interesting article on this topic by Lehmann, "The Dome of Heaven," which was reprinted in Kleinbauer, *Modern Perspectives,* pp. 227–270.

3. An excellent book that covers the entire series of miniatures and includes color reproductions of most of the paintings is Cazelles and Rathofer, *Illuminations of Heaven and Earth.*

4. Among the many skies that appear in *Les Trés Riches Heures,* about half are blue and empty and half show clouds without any heavenly bodies. One magnificent exception is *The Entombment,* which shows a twilight sky streaked with orange-tinted clouds and set aglow by the final rays of the setting Sun. The towers of Jerusalem, the rock walls of Calvary, and even a small boatman rowing in a river below are touched by the rosy light. The Sun appears on the horizon in flattened form, just as it does in nature. The main human figures, however, are shown as if at midday; they are unaffected by the sunlight and are entirely without shadows, as if cut from a separate image and pasted here. The painting is a remarkable example of a transition in artistic consciousness, combining elements of past and present, of earlier medieval didacticism (teaching the textual stories of holy writ) and later medieval witness (emphasizing the external, material world).

5. See Meiss, *French Painting.*

6. Recent examples of debates over Van Eyck can be found in Pächt, *Van Eyck;* Harbison, *Jan Van Eyck;* and Hall, *Arnolfini Betrothal.*

7. Reaves and Pedretti, "Leonardo da Vinci's Drawings."

8. Montgomery, "Naturalistic Drawings."

9. See, for example, Smart, *Renaissance and Mannerism,* pp. 31–33; and Pächt, *Van Eyck,* pp. 51, 186–188, 203–204.

10. Gedzelman, "The Sky in Art"; Simon, "Die ersten Aufschlüsse"; and Montgomery, "The Eye and the Rock." Gedzelman's fascinating article discusses Van Eyck's *The Crucifixion* and is worth quoting at length:

> This small work is the closest approach to a cloud atlas in the history of art. . . . It is even possible that the *Crucifixion* contains a weather forecast, [because] the sky gives many signs of having been recently cleared. The compass orientation of the painting is provided by the Biblical event and by the waning gibbous moon [that] appears in the lower right side of the sky. Since at this phase the moon is slightly more than 90° from the sun—which is off to the left—and the mid-afternoon early spring sun is in the southwest, the scene faces north. The sloping cirrus clouds indicate a jet stream wind from the left or west, while the mere presence of so many high and middle clouds suggests mid-level moisture. Three windmills all face the northeast, presumably establishing the surface wind direction. Combining these signs with the modest-size cumulus in the crisp, near-surface air suggests that a slowly moving cold front has just passed and cleared the sky. (p. 11)

The author notes that the position of the Moon does not conform to the waning gibbous phase. This discrepancy indicates that Van Eyck may have done a separate study of the Moon, as he apparently did of the mountain background, the limestone outcrop, and portions of the sky, all of which he evidently grafted together to produce the final scene.

11. This type of reading mostly originates from the work of the art historian, Erwin Panofsky. See especially "Arnolfini Portrait" and his much more extensive *Early Netherlandish Painting.*

12. Seidel, "Arnolfini Portrait." See also Seidel's more recent article, "Value of Verisimilitude." For recent arguments against this type of reading, the best critique is by Hall, *Arnolfini Betrothal.*

13. Dvořák, "Das Rätsel."

14. Panofsky, *Early Netherlandish Painting,* vol. 1, p. 8.

15. See Harbison, *Jan Van Eyck,* pp. 20–22. An interest in alchemy apparently launched Van Eyck on his experiments with oil paint, which dramatically raised the level of illusionistic effects that could be achieved, setting an entirely new standard.

16. Ibid., p. 24.

17. Panofsky, *Early Netherlandish Painting,* vol. 1, p. 182.

18. Richter, *Literary Remains,* p. 235.

19. Most of these writings have been collected in Richter, *Notebooks,* vol. 2; see pp. 154–168.

CHAPTER 7. THE BRITISH CONTRIBUTION

1. *De mundo* was later published in 1651 in Amsterdam. Kelly, *De mundo.*

2. Compare Whitaker, "Selenography" with Kopal and Carder, *Mapping of the Moon.* Of these two works, the essay by Whitaker (who argues that Gilbert's image is indeed a map) is considerably more accurate and informative on early lunar cartography.

3. Zilsel, "Gilbert's Scientific Method."

4. Ibid., pp. 36–37.

5. Ptolemy's text was compiled in A.D. 160, brought to Italy from Byzantium in 1406, translated into Latin shortly thereafter, and first published in the late 1470s with maps displaying more than eight thousand place names.

6. Zilsel, "Gilbert's Scientific Method," pp. 2–3.

7. See, for example, Johnson, *Astronomical Thought,* pp. 226–230; and Rosen, "Harriot's Science."

8. Quinn, "Thomas Harriot." More detail is provided in Shirley, *Thomas Harriot.*

9. Pepper, "Harriot's Earlier Work."

10. Harriot, *Briefe and True Report,* pp. 108–109.

11. See Pepper, "Harriot's Earlier Work."

12. Sluiter, "The Telescope before Galileo." See also the excellent study, Van Helden, "Invention of the Telescope."

13. Whitaker, "Selenography."

14. Bloom, "Borrowed Perceptions." There has been some debate over whether Harriot actually received a copy of Galileo's work or just heard about it, possibly via Kepler's *Dissertatio cum nuncio sidereo* (A conversation with Galileo's sidereal messenger), published in May 1610. At least one other English scientist, Sir Christopher Heydon, had procured a copy of Galileo's book by early July. Because of Harriot's general standing in England and his research in astronomy and optics, it seems likely that he would have been apprised of the new invention by this time. See North, "Thomas Harriot."

15. Shirley, "Thomas Harriot's Lunar Observations."

16. Ibid., p. 303.

17. Ibid.

18. Shirley, *Thomas Harriot,* p. 387.

19. Quoted in Whitaker, "Selenography," p. 120.

20. Ibid., pp. 120–121.

21. Several of these pages are reproduced in Shirley, "Thomas Harriot's Lunar Observations," pp. 291–300.

CHAPTER 8. GALILEO

1. Ariew, "Galileo's Lunar Observations," p. 214.

2. Kepler, *Astronomiae pars optica,* pp. 216–221. Interestingly, Kepler reminds Galileo of this and many other precedents to his discoveries in his *Dissertatio;* see p. 27 of the translation by Edward Rosen.

3. Kepler, *Astronomiae pars optica,* pp. 216–221.

4. Evidence for this can be found in Kepler's *Dissertatio.* Kepler notes, for example, that his own teacher, Michael Mästlin, had written a brief work in which much was said about the "kinship of the Moon with the earth on the basis of their density, shadow, atmosphere, and light borrowed from the sun" (p. 30). Mästlin apparently drew an image of the lunar surface during an eclipse, showing a large spot representing the shadow. This work is now unfortunately lost. But Kepler, being so reliable in everything else he cites, should be taken as an authority here.

5. See, for example, Drake, "Galileo's First Telescopic Observations"; Whitaker, "Selenography"; Westfall, "Scientific Patronage"; and Van Helden, "Galileo."

6. Galileo, *Sidereus nuncius,* p. 35.

7. Winkler and Van Helden, "Representing the Heavens."

8. Kepler, *Dissertatio,* p. 12.

9. Caspar, *Kepler,* p. 351.

10. Ibid., p. 352.

11. Such a comparison is presented in Whitaker, "Selenography," p. 125.

12. Galileo, *Sidereus nuncius,* pp. 51–53 and n. 47.

13. The various methods are discussed in Cajori, "History of Determinations."

14. Whitaker, "Selenography," p. 124.

15. Edgerton, *Giotto's Geometry.*

16. Viviani, "Racconto istorico," p. 602.

17. At one point in his discussion, Galileo mentions that the "cavity larger than all others . . . offers the same aspect to shadow and illumination as a region similar to Bohemia would offer on Earth, if it were enclosed on all sides by very high mountains, placed round the periphery in a perfect circle" (p. 47). Galileo's attempt here is rhetorical: he uses analogy to bolster his case that the Moon is like the Earth.

18. Drake, *Galileo Studies.*

19. The larger role of images in Galileo's book is a rich topic for investigation. In addition to engravings of the Moon, he included mathematical diagrams, sketches of several constellations with stars depicted symbolically, and sixty versions of Jupiter and its moons arranged horizontally and inserted directly into the text. In each of these cases, a different style of observation was displayed. The Moon, however, demanded a much greater degree of care and effort: Galileo apparently recognized the need to demonstrate his own witness of the lunar surface by providing a true picture that seemed to embody an actual perception.

20. Reeves, *Painting the Heavens.*

21. Van de Vyver, "Early Lunar Maps."

22. See Whitaker, "Selenography," and Winkler and Van Helden, "Representing the Heavens."

23. See Cavina, "On the Theme of Landscape." Cavina's basic information about *Flight into Egypt* and its use of Galilean observations is excellent. However, her insistence that "the sky [in this painting] is studied from nature, observed and recorded as it had never been up to then" (p. 140) is less tenable, given the achievements of Van Eyck, the Limbourg brothers, and Galileo himself.

24. This was surmised on the basis of the following: the Earth's rotation of 360 degrees in 24 hours translates into 0.25 degrees of longitude (approximately 17.3 miles at the equator) every minute. Therefore, if two observers at different locations could perceive the same phenomenon coming into view at the same moment and note the exact time they did so, the difference in these logged times could be converted into the distance between these observers.

25. Revel, "Knowledge of the Territory," p. 137.

CHAPTER 9. RETURN OF THE TEXT

1. Nicolson, *Voyages to the Moon* and *World in the Moon.*

2. Nicolson, *World in the Moon,* p. 2.

3. Ariosto, *Orlando furioso,* p. 294.

4. The court masque was a genre of entertainment, now defunct, that consisted of a few dozen pages of dialogue and verse commonly full of wit, puns, salacious asides, and topical references intended for the ears and eyes of the king and his retinue. Though brief in text, a masque would be spun out into an extravagant musical pageant that was hours in length and performed with considerable scenery and fanfare. Most often it was intended as a celebration and flattery of the court. For a good introduction to the masque, see the somewhat dated but still useful work, Welsford, *The Court Masque.*

5. Sellin, "The Performances of Ben Jonson's *Newes*."

6. See, for example, Orgel, *The Jonsonian Masque*.

7. Sellin, "The Politics of Ben Jonson's *Newes*."

8. Jonson, *The Complete Masques,* p. 293.

9. These are discussed in detail in Sellin, "The Politics of Ben Jonson's *Newes*."

10. Jonson, *Newes,* in *The Complete Masques,* p. 302.

11. Kepler, *Somnium,* p. 11. The "quarrel" was a challenge by the archduke for the emperorship. Libussa was an early medieval ruler of Bohemia whose authority was apparently challenged by a male uprising. See the notes, especially n. 14, provided by Rosen in his translation.

12. Ibid., p. 150, nn. 7, 8.

13. Godwin, *Man in the Moone.*

14. Douglas Bush mentions there were twenty-five editions in four languages by 1768. See his discussion of Godwin in *Oxford History.*

15. de Bergerac, *L'autre monde,* p. 66.

16. Ibid., pp. 72–73.

17. Ibid., p. 148.

18. For a detailed discussion of Wilkins's work, see Nicolson, *Voyages to the Moon,* pp. 1–32.

19. Wilkins, *Discovery of a World,* pp. 212–213.

20. Ibid., p. 11.

21. Quoted in Davy, *British Scientific Literature,* pp. 220–221.

22. Milton, *Paradise Lost,* p. 171.

CHAPTER 10. EFFORTS FROM FRANCE AND BELGIUM

1. See Humbert, "La première carte."

2. Quoted in Reeves, *Painting the Heavens,* p. 12.

3. Humbert, "La première carte," p. 198.

4. Ibid., p. 200.

5. See, for example, Bosmans, "La carte lunaire . . . à Bruxelles" and "La carte lunaire . . . à l'Université Leyde"; Prinz, "L'original"; and Wislicenus, "Ueber die Mondkarten." A more general placement of Van Langren in seventeenth-century attempts at mapping the Moon can be found in Whitaker, "Selenography."

6. Whitaker, "Selenography," p. 128.

7. Bosmans, "La carte lunaire . . . à Bruxelles," p. 113.

8. Van de Vyver, "Lettres," p. 83.

9. Galileo, *Sidereus nuncius,* pp. 29–31.

10. This is noted in Wislicenus, "Ueber die Mondkarten"; Prinz, "L'original"; and Bosmans, "La carte lunaire . . . à l'Université Leyde."

11. This represents the major part of Van Langren's inscription, translated from a French version of the original Latin provided in Bosmans, "La carte lunaire . . . à l'Université Leyde," pp. 251–254.

12. Van de Vyver, "Lettres," p. 86.

13. Bosmans, "La carte lunaire . . . à Bruxelles," pp. 123–124.

14. Ibid., p. 130.

CHAPTER 11. JOHANNES HEVELIUS

1. Quoted in Winkler and Van Helden, "Johannes Hevelius," p. 105.

2. All of these elements of Hevelius's observatory are noted in a contemporary account of the tragic fire in 1679 that destroyed the entire edifice, including a large number of unpublished papers. For a complete translation of the original letter in which this account appears, see MacPike, *Hevelius,* pp. 103–111.

3. Despite his importance to the history of astronomy, there is no modern full-length biography of Hevelius. Instead a portrait of his life and achievements must be gleaned from a range of sources: MacPike, *Hevelius;* North, "Johannes Hevelius"; Volkoff et al., *Johannes Hevelius;* Beziat, "La vie et les travaux"; Lengnich, *Hevelius;* Brandstäter, *Johannes Hevelius,* pp. 360–364; and Westphal, *Leben.* Hevelius himself gives an interesting autobiographical account of his scientific education under Krüger in the preface to his work, *Machinae coelestis pars prior.*

4. See Lengnich, *Hevelius,* pp. 1–30; and Brandstäter, *Johannes Hevelius,* chapters 1 and 2.

5. Quoted in Volkoff et al., *Johannes Hevelius,* p. 12.

6. Beziat, "La vie et les travaux," p. 504.

7. Quoted in Volkoff et al., *Johannes Hevelius,* p. 15. A German version of the entire letter can be found in Archenhold, "Johannes Hevelius."

8. Winkler and Van Helden, "Johannes Hevelius," p. 106.

9. Quoted in Wolf, *History of Science,* p. 168.

10. This point has been discussed briefly in Van de Vyver, "Lunar Maps," and in more detail in Govi, "Della invenzione." It is also noted by Whitaker in *Mapping and Naming the Moon.* The relevant Latin phrase is as follows: *sed lente vitrea subtilissimis filis adinstar craticulae dispositis operta, qua ipsas Lunae maculas delineauit, et suo quamque loco propria manu exactissimi posuit* (roughly translated as "very fine threads were placed on the lens and used to sketch the lunar spots, so that whatever their location, it could be drawn as precisely as possible").

11. I owe this distinction, which is rich with historical possibilities, to Professor A. Van Helden.

12. Hooke, *Animadversions.*

13. A brief but accurate description of the controversy can be found in Volkoff et al., *Johannes Hevelius*, pp. 36–46.

14. Although the use of *putti* on maps, architectural drawings, and other images was standard practice during the baroque period, Hevelius was the first to introduce them into seventeenth-century lunar imagery—the first, that is, to feel that they belonged there.

15. Despite his training and skill in this area, Hevelius did not draw or engrave this or several other images in his works. Instead he employed draftsmen to help perform this labor. Two of these, Adolph Boy and S. Falk, were responsible for the nonlunar pictures, whereas Hevelius drew and engraved all of the images of the Moon, which carry the attribution *Autor sculpsit* ("engraved by the author").

16. Hevelius, *Selenographia,* p. 223.

17. See, for example, Strobell and Masursky, "Planetary Nomenclature."

18. Hevelius, *Selenographia,* p. 353.

19. Ibid.

CHAPTER 12. RICCIOLI

1. The best sources on Riccioli include the following: Campedelli, "Riccioli, Giambattista"; Bailly, *Histoire,* vol. 2; Libes, *Histoire,* vol. 2; Delambre, *Histoire,* vol. 2; and Harris, "Jesuit Ideology."

2. Quoted in Nicolson, *World in the Moon.*

3. Campedelli, "Riccioli, Giambattista," p. 411.

4. Delambre, *Histoire,* vol. 2, p. 275.

5. Ibid., p. 279.

6. The sentence contains a pun, which is difficult to render correctly in English but translates loosely as, "I am straightened up at the same time that I am straightened out."

7. *Psalms* 8:4.

8. The relevant story of Jove and Io takes place at the end of Ovid's *Metamorphoses,* Book 1.

9. See, for example, Bailly, *Histoire,* and Libes, *Histoire.*

10. The most sympathetic dismissal I have found is offered by Libes, *Histoire,* who says of Riccioli, "One owes him some recognition . . . finally, for having worked constantly in a useful manner, if not for science at least for those who cultivate it" (p. 94).

11. See Van de Vyver, "Lunar Maps," p. 78.

12. Ibid.

13. See, for example, Whitaker, "Selenography," p. 138.

14. See Delambre, *Histoire,* p. 282. Note that this was written in the early nineteenth century.

15. For a discussion of this gradual disappearance, see Montgomery, "Naming the Heavens," pt. 2.

16. See Whitaker, *Mapping and Naming the Moon,* especially chapters 5 and 6.

17. Van de Vyver, "Lunar Maps."

18. Ibid., p. 80.

19. See Beziat, "La vie et les travaux," p. 506.

20. See Montgomery, *Scientific Voice,* chapter 5.

21. Much of the controversy is discussed in MacPike, *Hevelius.*

22. Schröter, *Selenotopografisches Fragmenten,* vol. 1, pp. 70–71.

CHAPTER 13. A LUNAR LEGACY

1. Quoted in Sheehan, *Worlds in the Sky,* p.155.

2. The most complete version can be found in Grosser, *Discovery of Neptune.* Another excellent source is Baum and Sheehan, *In Search of Planet Vulcan.* Good summaries also appear in Sheehan, *Worlds in the Sky,* chapter 12, and in North, *Astronomy and Cosmology,* pp. 427–430.

3. Sheehan, *Worlds in the Sky,* p. 176.

4. Quoted in Blunck, *Mars and Its Satellites,* p. 14.

5. Ibid., p. 16.

6. These words from J. B. Lindsay are recorded as part of a general session on the question of Martian nomenclature, published in volume 17 of the *The Astronomical Register* (1879), p. 25.

7. Blunck, *Mars and Its Satellites,* p. 20.

8. Schiaparelli, "Sulla rotazione." See also the brief review in Sheehan, *Worlds in the Sky,* pp. 52–55.

9. Similar points have been made in Sheehan, *Planets and Perception.*

10. An excellent reproduction of this image, which was published in a 1690 volume by J. Seller, titled *Atlas coelestis,* appears in Whitfield, *Mapping of the Heavens,* p. 108.

CHAPTER 14. THE LUNAR CYCLE

1. Elliot, *The Old World in the New,* p. 8.

2. Quoted in Bachelard, *La formation,* pp. 224–225.

3. Wilson, *Invisible World,* p. 251.

4. Bachelard, *La formation,* p. 5.

5. Blagg and Müller, *Named Lunar Formations.*

6. Whitaker et al., *Rectified Lunar Atlas.*

7. Firsoff, *Strange World of the Moon,* p. 172.

Bibliography

Albertus Magnus (Albert the Great). *Alberti Magni Opera omnia.* Edited by
    B. Geyer. Aschendorff: Monasterii Westfalorum, 1964.

al-Haytham, A. *Abhandlung des Schaichs ibn 'Ali al-Hasan ibn al-Hasan ibn al-
    Haitham: Über die Natur der Spuren (Flecken), die man auf der Oberfläche des Mondes
    sieht.* Translated by C. Schoy. Hannover: Heinz Lafaire, 1925.

Anonymous. *Theorica planetarum.* In *A Sourcebook in Medieval Science,* edited by
    E. Grant, 451 465. Cambridge, Mass.: Harvard University Press, 1974.

Aratus. *Phænomena.* Translated by G. R. Mair. Loeb Classical Library. Cambridge,
    Mass.: Harvard University Press, 1955.

Archenhold, F. S. "Johannes Hevelius." *Das Weltall: illustrierte Zeitschrift für Astron-
    omie und verwandte Gebiete* 11 (1911): 152–154.

Ariew, R. "Galileo's Lunar Observations in the Context of Medieval Lunar Theory."
    Studies in the History and Philosophy of Science 15, 3 (1984): 213–226.

Ariosto, L. *Orlando furioso.* 1532. Translated by Sir John Harington, London:
    Richard Field, 1607. Reprint (selections), edited by R. Gottfried, Bloom-
    ington, Ind.: Indiana University Press, 1963.

Aristotle, *De caelo* (On the heavens). Translated by W. K. C. Guthrie. Loeb Classical
    Library. Cambridge, Mass.: Harvard University Press, 1939.

——. *Generation of Animals.* Translated by A. L. Peck. Loeb Classical Library. Cam-
    bridge, Mass.: Harvard University Press, 1943.

——. *Meteorologica.* Translated by H. D. P. Lee. Loeb Classical Library. Cambridge,
    Mass.: Harvard University Press, 1952.

——. *Parts of Animals, Movement of Animals, Progression of Animals.* Translated by
    A. L. Peck and E. S. Forster. Loeb Classical Library. Cambridge, Mass.: Har-
    vard University Press, 1933.

Ashbrook, J. *The Astronomical Scrapbook.* Cambridge: Cambridge University Press, 1984.

Bachelard, G. *La formation de l'esprit scientifique.* 2d ed. Paris: J. Vrin, 1983.

Bailly, J. S. *Histoire de l'astronomie moderne.* 2 vols. Paris: DeBure, 1779.

Baum, R., and W. Sheehan. *In Search of Planet Vulcan: The Ghost in Newton's Clockwork Universe.* New York: Plenum, 1997.

Beckwith, J. *Early Medieval Art.* London: Thames & Hudson, 1969.

Bede, V. *Bedae Venerabilis Opera.* Edited by C. W. Jones. In *Corpus Christianorum, series Latina.* Turnholt: Brepols, 1967.

Benson, R. L., and G. Constable, eds. *Renaissance and Renewal in the Twelfth Century.* Cambridge, Mass.: Harvard University Press, 1982.

Beziat, L. C. "La vie et les travaux de Jean Hevelius." *Bullettino di bibliografia e di storia delle scienze matematiche e fisiche* 8 (1875): 497–558, 589–669.

Biancani, G. *Sphaera mundi.* Bononi: Typis Sebastiani Bonomij, sumptibus Hieronymi Tamburini, 1620. Facsimile reprint, New York: Readex Microprint Corporation, 1974.

Blaauw, A. *History of the* IAU: The Birth and First Half Century of the International Astronomical Union. Dordrecht: Kluwer Academic Publishers, 1994.

Blagg, M., and K. Müller. *Named Lunar Formations.* London: Percy Lund & Humphries, 1935.

Bloom, T. "Borrowed Perceptions: Harriot's Maps of the Moon." *Journal for the History of Astronomy* 9 (1978): 117–122.

Blunck, J. *Mars and Its Satellites: A Detailed Commentary on the Nomenclature.* Hicksville, N.Y.: Exposition Press, 1977.

Bosmans, H. "La carte lunaire de Van Langren conservée à l'Université Leyde." *Revue des questions scientifiques,* 3rd ser., 17 (1910): 248–264.

——. "La carte lunaire de Van Langren conservée aux archives général du royaume, à Bruxelles." *Revue des questions scientifiques,* 3rd ser., 4 (1903): 108–139.

Bouché-Leclerc, A. *L'astrologie grecque.* Paris: Leroux, 1899.

Brandstäter, F. A. *Johannes Hevelius sein Leben und seine Bedeutsamkeit.* Danzig: J. A. Weber, 1861.

Brueton, Diana. *The Moon: Myth, Magic and Fact.* London: Labrynth Publishing, 1998.

Bush, D. *Oxford History of English Literature: The Early Seventeenth Century.* Oxford: Oxford University Press, 1945.

Cajori, F. "History of Determinations of the Heights of Mountains." *Isis* 12 (1929): 482–514.

Calcidius. *Timaeus a Calcidio translatus commentarioque instructus* (Commentary on

Plato's Timaeus). Edited by J. H. Waszink and P. J. Jensen. Vol. 4 of *Plato Latinus,* edited by R. Klibansky. London: Warburg Institute, 1962.

Campedelli, L. "Riccioli, Giambattista." In *Dictionary of Scientific Biography.* Vol. 9, 411–412. New York: Scribner's, 1975.

Caspar, M. *Kepler.* Rev. ed. New York: Dover, 1993.

Cavina, A. O. "On the Theme of Landscape—II: Elsheimer and Galileo." *Burlington Magazine* 118 (1976): 139–144.

Cazelles, R., and J. Rathofer. *Illuminations of Heaven and Earth: The Glories of the Trés Riches Heures du Duc de Berry.* New York: Harry N. Abrams, 1988.

Clagget, M. *Greek Science in Antiquity.* London: Abelard-Schulman, 1957. Reprint, New York: Barnes & Noble, 1994.

Clark, K. *Landscape into Art.* Boston: Beacon Press, 1961.

Classen, J. "Die Ältesten Mondkarte." *Jahrgang* 22, Heft 1 (1942): 1–16.

Dante Aligheri. *The Convivio of Dante Aligheri.* Translated by P. H. Wicksteed. London: J. M. Dent, 1908.

——. *The Divine Comedy.* Translated by J. A. Carlyle, T. Okey, and P. H. Wicksteed. New York: Random House, 1932.

Davy, N., ed. *British Scientific Literature in the Seventeenth Century.* London: George G. Harrap, 1953.

de Bergerac, C. *L'autre monde: Les etats et empires de la lune; les etats et empires du soleil.* 1649. Paris: Éditions Sociales, 1978.

de Bruyne, E. *L'Esthétique du moyen age.* Louvain: Éditions de l'Institut Supéreur de Philosophie, 1947.

Delambre, J. B. *Histoire de l'astronomie moderne.* 1821. Reprint (2 vols.), New York: Johnson Reprint Corp, 1969.

Dhanens, E. *Hubert and Jan Van Eyck.* New York: Konigstein im Taunus, 1980.

Dicks, D. R. *Early Greek Astronomy to Aristotle.* Ithaca, N.Y.: Cornell University Press, 1970.

Diels, Hermann. *Die Fragmente die Vorsokratiker.* 5th ed. Berlin: Weidmannsche Buchhandlung, 1934.

Dilke, O. A. W. *Greek and Roman Maps.* Ithaca, N.Y.: Cornell University Press, 1985.

Diogenes Laertius. *Lives of the Eminent Philosophers.* 2 vols. Translated by R. D. Hicks. Loeb Classical Library. Cambridge, Mass.: Harvard University Press, 1925.

Diringer, D. *The Illuminated Book: Its History and Production.* New York: Praeger, 1967.

Dowden, K., trans. *The Alexander Romance.* In *Collected Ancient Greek Novels,* edited by B. P. Reardon. Berkeley: University of California Press, 1989.

Drake, S. "Galileo's First Telescopic Observations." *Journal for the History of Astronomy* 7 (1976): 153–168.

———. *Galileo Studies.* Ann Arbor: University of Michigan Press, 1970.

Dreyer, J. L. E. *A History of Astronomy from Thales to Kepler.* New York: Dover, 1953.

Dvořák, M. "Das Rätsel der Kunst der Brüder van Eyck." *Jahrbuch der Kunsthistorischen Sammlung in Wien* 24, Heft 5 (1904): 161–204.

———. "Idealismus und Naturalismus in der gotischen Skulptur und Malerei." *Historische Zeitschrift* 119 (1918): 1–62, 185–246.

Eastwood, B. "The Astronomies of Pliny, Martianus Capella, and Isidore of Seville in the Carolingian World." In *Science in Western and Eastern Civilization in Carolingian Times,* edited by P. L. Butzer and D. Lohrmann, 161–180. Basel: Birkhaüser, 1993.

———. "Astronomy in Christian Latin Europe c. 500–1150 A.D.." *Journal for the History of Astronomy* 28 (1997): 235–258.

———. "Origins and Contents of the Leiden Planetary Configuration: An Artistic Schema of the Early Middle Ages." *Viator* 14 (1983): 1–40.

———. "Plinian Astronomical Diagrams in the Early Middle Ages." In *Mathematics and Its Applications to Science and Natural Philosophy in the Middle Ages,* edited by E. Grant and J. Murdoch, 141–172. Cambridge: Cambridge University Press, 1987.

Eco, U. *Art and Beauty in the Middle Ages.* New Haven: Yale University Press, 1986.

Edgerton, S. Y., Jr. *The Heritage of Giotto's Geometry: Art and Science on the Eve of the Scientific Revolution.* Ithaca, N.Y.: Cornell University Press, 1991.

Elliot, J. H. *The Old World in the New.* Cambridge: Cambridge University Press, 1970.

Evelyn-White, H. G., trans. *The Homeric Hymns and Homerica.* Loeb Classical Library. Cambridge, Mass.: Harvard University Press, 1914.

Firsoff, V. A. *Strange World of the Moon.* New York: Science Editions, 1962.

Freeman, K. *The Presocratic Philosophers.* Cambridge, Mass.: Harvard University Press, 1966.

Galileo Galilei. *Sidereus nuncius.* 1610. Translated by A. Van Helden. Chicago: University of Chicago Press, 1989.

Ganzenmüller, W. *Das Naturgefühl im Mittelalter.* Leipzig: B. G. Teubner, 1914.

Gedzelman, S. D. "The Sky in Art." *Weatherwise* (December 1991/January 1992): 8–12.

Gimpel, J. *The Medieval Machine.* New York: Penguin, 1983.

Godwin, F. *The Man in the Moone* (1638) and *Nuncius inanimatus* (1635). Edited by G. McColley. Smith College Studies in Modern Languages 19 (1937).

Goetz, W. "Die Entwicklung des Wirlichkeitssinnes vom 12. Zum 14. Jahrhundert." *Archiv für Kulturgeschichte* 27 (1937): 33–73.

Gombrich, E. H. *Art and Illusion.* Princeton, N.J.: Princeton University Press, 1961.

Goodenough, J. E. R. *Jewish Symbols in the Greco-Roman Period.* 13 vols. New York: Bollingen, 1964.

Govi, G. "Della invenzione del micrometro per gli strumenti astronomici." *Bullettino di Bibliografia di Storia delle Scienze Matematiche e Fisiche* 20 (1887): 607–622.

Grant, E., ed. *A Sourcebook in Medieval Science.* Cambridge, Mass.: Harvard University Press, 1974.

Grosser, M. *The Discovery of Neptune.* New York: Dover, 1979.

Guiley, R. *The Lunar Almanac.* London: Judy Piatkus, 1991.

Guthrie, W. K. C. *A History of Greek Philosophy.* 3 vols. Cambridge: Cambridge University Press, 1962, 1965, 1978.

——. *Orpheus and Greek Religion.* London: Methuen & Co., 1935. Reprint, Princeton, N.J.: Princeton University Press, 1993.

Hall, E. *The Arnolfini Betrothal: Medieval Marriage and the Enigma of Van Eyck's Double Portrait.* Berkeley: University of California Press, 1994.

Harbison, C. *Jan Van Eyck: The Play of Realism.* London: Reaktion Books, 1991.

Harriot, T. *A Briefe and True Report of the New Found Land of Virginia.* 1588. In *Voyages to the Virginia Colonies,* edited by A. L. Rowse, 107–136. London: Century, 1986.

Harris, S. "Jesuit Ideology and Jesuit Science: Scientific Activity in the Society of Jesus, 1540–1773." Ph.D. diss., University of Wisconsin, 1988.

Harris, W. V. *Ancient Literacy.* Cambridge, Mass.: Harvard University Press, 1989.

Haskins, C. H. *The Renaissance of the Twelfth Century.* Cambridge, Mass.: Harvard University Press, 1927.

Havelock, E. "The Linguistic Task of the Presocratics." In *Language and Thought in Early Greek Philosophy,* edited by K. Robb, 7–82. La Salle, Ill.: Hegeler Institute, 1983.

Heath, T. L. *Aristarchus of Samos.* Oxford: Oxford University Press, 1912.

——. *Greek Astronomy.* London: J. M. Dent & Sons, 1932.

Hesiod. *The Homeric Hymns and Homerica.* Translated by H. G. Evelyn-White. Loeb Classical Library. Cambridge, Mass.: Harvard University Press, 1914.

Hevelius, J. *Machinae coelestis pars prior.* Danzig: S. Reiniger, 1673–1679. Facsimile reprint, New York: Readex Microprint Corporation, 1981.

——. *Selenographia: sive, Lunae Descriptio; atque Accurata, tam Macularum ejus, quam Motuum Diversorum, etc.* 1647. New York: Johnson Reprint Co., 1967.

Hollister, C. W., ed. *The Twelfth Century Renaissance*. New York: Wiley, 1969.

Hooke, R. *Animadversions on the Machina Coelestis of the Honourable, Learned, and deservedly Famous Astronomer Johannes Hevelius, consul of Dantzick; Together with an Explication of some Instruments*. London: John Martin, 1674.

Horrox, J. "The Transit of Venus." 1640. Reprinted in Nicolson, M. H., *A World in the Moon: A Study of the Changing Attitude toward the Moon in the Seventeenth and Eighteenth Centuries*. Smith College Studies in Modern Languages 17 (1935): 23.

Humbert, P. "La première carte de la lune." *Revue des questions scientifiques* 20 (1931): 193–204.

Johnson, F. R. *Astronomical Thought in Renaissance England*. New York: Octagon Books, 1968.

Jones, H. S. *John Couch Adams and the Discovery of Neptune*. Cambridge: Cambridge University Press, 1947.

Jonson, B. *The Complete Masques*. Edited by Stephen Orgel. New Haven: Yale University Press, 1969.

Katzenstein, R., and E. Savage-Smith. *The Leiden Aratea: Ancient Constellations in a Medieval Manuscript*. Malibu, Calif.: J. Paul Getty Museum, 1988.

Kelly, S. *The De mundo of William Gilbert*. Amsterdam: Menno Hertzberger, 1965.

Kepler, J. *Astronomiae pars optica*. 1604. Vol. 2 of *Gesammelte werke*. Edited by W. von Dyck and M. Caspar. Munich: C. H. Beck, 1937.

——. *Dissertatio cum nuncio siderio* (A conversation with Galileo's sidereal messenger). 1610. Reprint, translated by Edward Rosen, New York: Johnson Reprint Corporation, 1965.

——. *Gesammelte werke* (Collected works). Reprint (20 vols.), edited by F. Hammer, Munich: C. H. Beck, 1937–1939.

——. *Somnium: The Dream, or Posthumous Work on Lunar Astronomy*. 1634. Translated by Edward Rosen, Madison: University of Wisconsin Press, 1967.

Kerenyi, K. *Die Mythologie der Griechen*. Zürich: Rhein Verlag, 1951.

Kern, O. *Orphicorum fragmenta*. Berlin: Weidmann, 1922.

Kirk, G. S., J. E. Raven, and M. Schofield. *The Presocratic Philosophers*. 2d ed. Cambridge: Cambridge University Press, 1983.

Kleinbauer, W. E., ed. *Modern Perspectives in Western Art History*. Toronto: University of Toronto Press, 1989.

Kopal, Z., and R. W. Carder. *Mapping of the Moon: Past and Present*. Dordrecht: D. Reidel, 1974.

Larsen, A. D. *Johannes Hevelius and His Catalog of Stars*. Provo, Utah: Brigham Young University Press, 1971.

Lehmann, K. "The Dome of Heaven." *Art Bulletin* 27 (1945): 1–27.

Lengnich, K. B. *Hevelius: oder Anekdoten und Nachrichten zur Geschichte dieses grossen Mannes: in Briefen, mit erläuternden Zusätzen und Beylagen.* Danzig: J. H. Florke, 1780.

Libes, A. *Histoire philosophique des progrès de la physique.* 2 vols. Paris: Immerzeel, 1812.

Lindberg, D. C. "The Transmission of Greek and Arabic Learning to the West." In *Science in the Middle Ages,* edited by D. C. Lindberg, 52–90. Chicago: University of Chicago Press, 1978.

Livingstone, D. N. *The Geographical Tradition.* Oxford: Basil Blackwell, 1992.

Lloyd, G. E. R. *Early Greek Science: Thales to Aristotle.* New York: W. W. Norton, 1970.

———. *The Revolutions of Wisdom: Studies in the Claims and Practice of Ancient Greek Science.* Berkeley: University of California Press, 1987.

Lohne, J. "Thomas Harriot." In *Dictionary of Scientific Biography.* Vol. 6, 124–129. New York: Scribner's, 1972.

Long, P. O., ed. *Science and Technology in Medieval Society.* Annals of the New York Academy of Sciences 441 (1985).

Lucian. *Icaromenippus.* Translated by A. M. Harmon. Loeb Classical Library. Cambridge, Mass.: Harvard University Press, 1915.

———. *A True Story.* Translated by A. M. Harmon. Loeb Classical Library. Cambridge, Mass.: Harvard University Press, 1913.

MacPike, E. F. *Hevelius, Flamsteed, and Halley.* London: Taylor & Francis, 1937.

Macrobius. *Commentary on the Dream of Scipio.* Translated by W. Stahl. New York: Columbia University Press, 1952.

Mâle, E. *The Gothic Image: Religious Art in France of the Thirteenth Century.* Translated by D. Nussey. New York: Harper, 1958.

Martianus Capella. See Stahl et al., 1972, 1977.

Martin, J. *Histoire du texte des Phenomenes d'Aratos.* Paris: C. Klincksieck, 1956.

McKitterick, R. "Text and Image in the Carolingian World." In *The Uses of Literacy in Early Mediaeval Europe,* edited by R. McKitterick, 297–318. Cambridge: Cambridge University Press, 1990.

Meiss, M. *French Painting in the Time of Jean de Berry: The Late Fourteenth Century and the Patronage of the Duke de Berry.* 2d ed. 2 vols. New York: Scribner's, 1969.

Milton, J. *Paradise Lost.* 1667. Reprint, edited by Scott Elledge, New York: Norton, 1975.

Montgomery, S. L. "A Brief History of Earthly Projections, Part 2: Nativizing Greco-Arabic Science." *Science as Culture* 6, 1 (1997): 73–129.

——. "The Eye and the Rock: Art, Observation, and the Naturalistic Drawing of Earth Strata." *Earth Sciences History* 15, 1 (1996): 3–24.

——. "The First Naturalistic Drawings of the Moon: Jan Van Eyck and the Art of Observation." *Journal for the History of Astronomy* 25 (1994): 317–320.

——. "Naming the Heavens: A Brief History of Earthly Projections." Pt. 1 and Pt. 2. *Science as Culture* 25 (1996): 546–587; 26 (1996): 73–129.

——. *The Scientific Voice.* New York: Guilford Press, 1996.

Moore, P. *The New Atlas of the Universe.* New York: Arch Cape Press, 1988.

Murdoch, J. *Album of Science: Antiquity and the Middle Ages.* New York: Scribner's, 1984.

Neckam, Alexander of. *De naturis rerum libri duo.* Edited by T. Wright. Nedeln, Liechtenstein: Kraus, 1967.

Neugebauer, O. *The Exact Sciences in Antiquity.* 2d ed. New York: Dover, 1969.

——. *A History of Ancient Mathematical Astronomy.* New York: Springer-Verlag, 1975.

Nicolson, M. H. *Voyages to the Moon.* New York: Macmillan, 1948.

——. *A World in the Moon: A Study of the Changing Attitude toward the Moon in the Seventeenth and Eighteenth Centuries.* Smith College Studies in Modern Languages 17 (1935): 1–71.

North, J. "Johannes Hevelius." In *Dictionary of Scientific Biography.* Vol. 6, 360–364. New York: Scribner's, 1972.

——. *The Norton History of Astronomy and Cosmology.* New York: Norton, 1995.

——. "Thomas Harriot and the First Telescopic Observations of Sunspots." In *Thomas Harriot, Renaissance Scientist,* edited by J. W. Shirley, 129–165. Oxford: Clarendon Press, 1974.

O'Leary, D. L. *How Greek Science Passed to the Arabs.* London: Routledge, 1949.

O'Neil, W. M. *Early Astronomy from Babylonia to Copernicus.* Sydney: Sydney University Press, 1986.

Oresme, N. *Le Livre du ciel et du monde.* 1377. Translated by A. D. Menut. Madison: University of Wisconsin Press, 1968.

Orgel, S. *The Jonsonian Masque.* Cambridge, Mass.: Harvard University Press, 1965.

Ortelius, A. *Theatrum orbis terrarum.* 1570. Reprint, edited by R. A. Skelton. Amsterdam: N. Israel, 1964.

Osborne, C. *Rethinking Early Greek Philosophy.* London: Duckworth, 1987.

Ovid. *Metamorphoses.* Translated by Rolfe Humphries. Bloomington, Ind.: Indiana University Press, 1955.

Pächt, O. *Van Eyck and the Founders of Early Netherlandish Painting.* London: Harvey Miller, 1994.

Panofsky, E. *Early Netherlandish Painting: Its Origins and Character.* 2 vols. Cambridge, Mass.: Harvard University Press, 1953.

———. "Jan van Eyck's Arnolfini Portrait." *The Burlington Magazine* 64 (1934): 117–127.

———. *Renaissance and Renascences in Western Art.* New York: Harper & Row, 1969.

Pepper, J. V. "Harriot's Earlier Work on Mathematical Navigation: Theory and Practice." In *Thomas Harriot, Renaissance Scientist,* edited by J. W. Shirley, 54–90. Oxford: Clarendon Press, 1974.

Plato. *The Collected Dialogues of Plato.* Edited by E. Hamilton and H. Cairns. Princeton, N.J.: Princeton University Press, 1963.

Pliny the Elder. *Naturalis historia* (Natural history). 12 vols. Translated by H. Rackam. Loeb Classical Library. Cambridge, Mass.: Harvard University Press, 1938–1955.

Plutarch. *On the Face Which Appears in the Orb of the Moon.* In *Plutarch's Moralia,* vol. 12, translated by H. Cherniss and W. Helmbold, 2–226. Loeb Classical Library. Cambridge, Mass.: Harvard University Press, 1957.

Prinz, W. "L'original de la première carte lunaire de Van Langren." *Ciel et Terre* 24 (1903–1904): 99–105, 149–155.

Ptolemy (Claudius Ptolemius). *Almagest.* Translated by G. J. Toomer. New York: Springer-Verlag, 1984.

———. *Geographia* (The geography). Translated and edited by E. L. Stevenson. London: Constable, 1991.

———. *Hypotheses of the Planets (Planetary Hypothesis).* In *Opera quae exstant omnia,* vol. 2 of *Opera astronomica minora,* edited by J. L. Heiberg, 69–145. Leipzig: Teubner, 1907.

———. *Tetrabiblos.* Translated by F. E. Robbins. Loeb Classical Library. Cambridge, Mass.: Harvard University Press, 1948.

Quinn, D. B. "Thomas Harriot and the New World." In *Thomas Harriot, Renaissance Scientist,* edited by J. W. Shirley, 36–53. Oxford: Clarendon Press, 1974.

Reaves, G., and C. Pedretti. "Leonardo da Vinci's Drawings of the Surface Features of the Moon." *Journal for the History of Astronomy* 18 (1987): 55–58.

Reeves, E. *Painting the Heavens: Art and Science in the Age of Galileo.* Princeton, N.J.: Princeton University Press, 1997.

Revel, J. "Knowledge of the Territory." *Science in Context* 4 (1991): 133–161.

Riccioli, G. *Almagestum novum astronomiam veterem novamque.* Bologna: Ex typograpia haeredis Victorii Benatii, 1651.

Richter, J. P. *The Literary Remains of Leonardo da Vinci.* 3rd ed. 2 vols. London: Sampson Low, Marston & Co., 1970.

——. *The Notebooks of Leonardo da Vinci.* 2 vols. New York: Dover, 1970.

Romm, J. S. *The Edges of the Earth in Ancient Thought.* Princeton, N.J.: Princeton University Press, 1992.

Rosen, E. "Harriot's Science, the Intellectual Background." In *Thomas Harriot, Renaissance Scientist,* edited by J. W. Shirley, 1–15. Oxford: Clarendon Press, 1974.

Sabra, A. I. "Ibn Al-Haytham." In *Dictionary of Scientific Biography* Vol. 6, 189–210. New York: Scribner's, 1972.

Sacrobosco, John of. *The Sphere of Sacrobosco and Its Commentators.* Edited and translated by L. Thorndike. Chicago: University of Chicago Press, 1949.

Sarton, G. *Hellenistic Science and Culture in the Last Three Centuries* B.C. New York: W. W. Norton, 1970.

Scheiner, C. *Disquisitiones mathematica.* Innsbruck: Daniel Agricolam, 1614.

Schiaparelli, G. V. "Sulla rotazione della Mercurio." *Astronomische Nachtrichten* 2944 (1889): 57–122.

Schoy, C. "Introduction." In *Abhandlung des Schaichs ibn Ali al-Hasan ibn al-Hasan ibn al-Haitham: Über die Natur der Spuren (Flecken), die man auf der Oberfläche des Mondes sieht,* translated by C. Schoy, vii–viii. Hannover: Heinz Lafaire, 1925.

Schröter, J. *Selenotopografisches Fragmenten zur genauern Kenntniss der Mondfläche.* 3 vols. Göttingen: Joh. Georg Rosenbusch, 1791.

Seidel, L. "Jan van Eyck's 'Arnolfini Portrait': Business as Usual?" *Critical Inquiry* 16 (1989): 55–86.

——. "The Value of Verisimilitude in the Art of Jan van Eyck." In *Contexts: Style and Values in Medieval Art and Literature,* edited by D. Poirion and N. F. Regalado, 25–43. New Haven: Yale University Press, 1991.

Seller, J. *Atlas coelestis, containing the systems and theoryes of the planets, the constellations of the starrs, and other phenomina's of the heavens, with nessesary tables relateing thereto.* London: Benjamin Bragg, 1690.

Sellin, P. R. "The Performances of Ben Jonson's *Newes from the New World Discover'd in the Moone.*" *English Studies* 61 (1980): 491–497.

——. "The Politics of Ben Jonson's *Newes from the New World Discover'd in the Moone.*" *Viator* 17 (1986): 321–338.

Sheehan, W. *Planets and Perception: Telescopic Views and Interpretations, 1609–1909.* Tucson: University of Arizona Press, 1988.

——. *Worlds in the Sky: The Story of Planetary Discovery from Earliest Times through Voyager and Magellan.* Tucson: University of Arizona Press, 1992.

Shirley, J. W. *Thomas Harriot: A Biography.* Oxford: Clarendon Press, 1983.

———. "Thomas Harriot's Lunar Observations." In *Science and History: Studies in Honor of Edward Rosen.* Studia Copernicana 16 (1978): 283–308.

Simon, W. "Die ersten Aufschlüsse im Bild und der Beginn der Geologie in Europa (1430–1550)." *Aufschluss* 27 (1976): 37–51.

Skelton, R. A. *Maps: A Historical Survey of Their Study and Collecting.* Chicago: University of Chicago Press, 1972.

Sluiter, E. "The Telescope before Galileo." *Journal of the History of Astronomy* 28 (1997): 223–235.

Smart, A. *The Renaissance and Mannerism in Northern Europe and Spain.* New York: Harcourt Brace Jovanovich, 1972.

Stahl, W. H. *Roman Science: Origins, Development, and Influence on the Later Middle Ages.* Madison: University of Wisconsin Press, 1962.

Stahl, W. H., R. Johnson, and E. L. Burge. *Martianus and the Seven Liberal Arts.* 2 vols. New York: Columbia University Press, 1972, 1977.

Stephens, S. A., and J. J. Winkler, eds. *Ancient Greek Novels: The Fragments.* Princeton, N.J.: Princeton University Press, 1995.

Stock, B. *The Implications of Literacy: Written Language and Models of Interpretation in the Eleventh and Twelfth Centuries.* Princeton, N.J.: Princeton University Press, 1983.

———. "Science, Technology, and Economic Progress in the Early Middle Ages." In *Science in the Middle Ages,* edited by D. C. Lindberg, 1–51. Chicago: University of Chicago Press, 1978.

Stott, C. *Celestial Charts: Antique Maps of the Heavens.* New York: Crescent, 1991.

Strobell, M. E., and H. Masursky. "Planetary Nomenclature." In *Planetary Mapping,* edited by R. Greeley and R. Bateson, 96–140. Cambridge: Cambridge University Press, 1990.

Summers, D. *The Judgement of Sense: Renaissance Naturalism and the Rise of Aesthetics.* Cambridge: Cambridge University Press, 1987.

Tatarkiewicz, W. *History of Aesthetics, Vol. 2: Medieval Aesthetics.* The Hague: Mouton, 1970.

Thorndike, L. *A History of Magic and Experimental Science.* 8 vols. New York: Columbia University Press, 1923.

Van de Vyver, O. "Lettres de J.-Ch. Della Faille, S.I., Cosmographe du Roi à Madrid, à M. F. Van Langren, Cosmographe du Roi à Bruxelles, 1634–1645." *Archivum Historicum Societatis Iesu* 46 (1977): 73–183.

———. "Lunar Maps of the XVIIth Century." *Vatican Observatory Publications* 1/2 (1971): 71–83.

——. "Original Sources of Some Early Lunar Maps." *Journal for the History of Astronomy* 7 (1971): 86–97.

Van Helden, A. "Galileo, Telescopic Astronomy, and the Copernican System." In *Planetary Astronomy from the Renaissance to the Rise of Astrophysics, Part A: Tycho Brahe to Newton,* edited by R. Taton and C. Wilson, 81–105. Cambridge: Cambridge University Press, 1989.

——. "The Invention of the Telescope." *Transactions of the American Philosophical Society* 67, pt. 4 (1977): 5–16.

Verdet, J.-P. *The Sky: Mystery, Magic, and Myth.* New York: Harry N. Abrams, 1992.

Viviani, V. "Racconto istorico." In Galileo Galilei, *Le Opere,* edited by A. Favaro, vol. 19, 597–632. Florence: Guinti Barbera, 1965.

Volkoff, I., E. Franzgrote, and A. D. Larsen. *Johannes Hevelius and His Catalog of Stars.* Provo, Utah: Brigham Young University Press, 1971.

von Euw, A. "Die Künstlerische Gestaltung der astronomischen und komputistischen Handschriften des Westens." In *Science in Western and Eastern Civilization in Carolingian Times,* edited by P. L. Butzer and D. Lohrmann, 251–269. Basel: Birkhaüser, 1993.

Weitzmann, K. *Late Antique and Early Christian Book Illumination.* New York: George Braziller, 1977.

Welsford, E. *The Court Masque.* Cambridge: Cambridge University Press, 1927.

Westfall, R. S. "Scientific Patronage: Galileo and the Telescope." *Isis* 76 (1985): 11–31.

Westphal, J. H. *Leben, Studien, und Schriften des Astronomen Johann Hevelius.* Koenigsberg: Hallervord, 1820.

Wheelwright, P. *The Presocratics.* New York: Macmillan, 1985.

Whitaker, E. A. *Mapping and Naming the Moon.* Cambridge: Cambridge University Press, 1999.

——. "Selenography in the Seventeenth Century." In *Planetary Astronomy from the Renaissance to the Rise of Astrophysics, Part A: Tycho Brahe to Newton,* edited by R. Taton and C. Wilson, 119–143. Cambridge: Cambridge University Press, 1989.

Whitaker, E. A., G. P. Kuiper, W. K. Hartmann, and H. L. Spradley. *The Rectified Lunar Atlas.* Tucson: University of Arizona Press, 1963.

White, L., Jr. *Medieval Technology and Social Change.* Oxford: Oxford University Press, 1962.

——. "Natural Science and Naturalistic Art in the Middle Ages." *The American Historical Review* 52, 3 (1947): 421–435.

Whitfield, P. *The Mapping of the Heavens.* San Francisco: Pomegranate, 1995.

Whyte, L., Jr. *Medieval Technology and Social Change.* London: Oxford University Press, 1962.

Wilkins, J. *The Discovery of a World in the Moone.* 1638. Reprint, New York: Scholars' Facsimiles & Reprints, Inc., 1973.

Williams, W. C. *Collected Poems: Volume 1, 1909–1939.* Edited by A. W. Litz and C. MacGowan. New York: New Directions, 1986.

Wilson, C. *The Invisible World: Early Modern Philosophy and the Invention of the Microscope.* Princeton, N.J.: Princeton University Press, 1995.

Winkler, M. G., and A. Van Helden. "Johannes Hevelius and the Visual Language of Astronomy." In *Renaissance and Revolution: Humanists, Scholars, Craftsmen, and Natural Philosophers in Early Modern Europe,* edited by J. V. Field and F. A. L. James, 97–116. Cambridge: Cambridge University Press, 1993.

———. "Representing the Heavens: Galileo and Visual Astronomy." *Isis* 83 (1992): 195–217.

Wislicenus, W., "Ueber die Mondkarten des Langrenus," *Biblioteca Mathematica,* 3rd ser., 2 (1901): 384–391.

Wolf, A. *A History of Science, Technology, and Philosophy in the Sixteenth and Seventeenth Centuries.* 2 vols. New York: Harper & Row, 1959.

Wright, H., H. Wright, and S. Rapport. *To the Moon.* New York: Meredith Press, 1968.

Yourcenar, M. *Mémoires d'Hadrien.* Paris: Plon, 1958.

Zilsel, E. "The Origins of William Gilbert's Scientific Method." *Journal of the History of Ideas* 2, 1 (1942): 1–53.

Index

Martianus Capella, 201; Oceanus Procellarum, 35, 200–201; Riccioli, 203; Theophilus, 202; Tycho, 156, 189; Van Langren, 202

lunar features, naming schemes for: Gassendi, 155–156; Gilbert, 100–101; Hevelius, 184–190; Riccioli, 198–208; Van Langren, 167

lunar surface, characteristics of: "roughness," 74–75; "spottedness," 42, 56, 65, 69, 76–79; visibility to naked eye, 227 n. 1

lunar surface, descriptions of: Albert the Great, 70; Al-Haytham, 71–76; Martianus Capella, 46; medieval views, 69–71; Plutarch's dialogue, 32–36

lunar surface, representations of: artists' renderings, 152–155; Galileo's, 114–120, 123–128; Harriot's, 106–113; Hevelius's, 172–174, 181–184; Leonardo's, 87, 95–97; post-Galilean drawings, 128–132; Van Eyck's, 87–95

lunar voyage, as literary theme, 135–150; Ariosto, 136–138; Cyrano de Bergerac, 146–148; Godwin, 143–146; Jonson, 138–140; Kepler, 140–143; Lucian, 36–40

Machinae coelestis pars prior (Hevelius), 171–172

Macrobius, *Commentary on the Dream of Scipio,* 45

"maculas," 42

Mädler, Johann, 212

Malapert, Charles, 130, 132, 193

man in the Moon, 65–69

Man in the Moone (Godwin), 144–146

mapmaking, 101, 124–128, 133–134, 163

mapping: of ancient world, 229 n. 40; of Solar System, 4–6, 58–61

mapping of Moon, 7–9, 220; Gilbert, 98–104; Hevelius, 172–174, 181–184; and longitude problem, 133–134; Peiresc-Gassendi project, 151–156; Riccioli, 198–208; Van Langren, 157–168

Marduk, 12

maria, lunar, 70, 89. *See also* lunar features

Marriage of Philology and Mercury (Martianus Capella), 45–46, 52

Mars, 212–216

Martianus Capella, *Marriage of Philology and Mercury,* 45–46, 52

Masaccio, *Acts of the Apostles,* 83

masculinity of Moon, 67

masque, 138–140, 237 n. 4

Mästlin, Michael, 236 n. 4

Mayer, Simon, 210

Mayer, Tobias, 181

Medicea Sidera (Medici stars), 127–128, 161–162, 210

Medusa, 14

Mellan, Claude, 153–155, 173

Mercator, Gerard, 105, 127

Mercator projection, 4–5

Mercury, 216–217

meteor, at Aegospotami, 20–21

Meteorologica (Aristotle), 27

Michelet, Jules, 222

microscope, 221–222

Milton, John, *Paradise Lost,* 150

"miracle letter," 31

moistness of Moon, 46, 95–97

Moon: Mercury and, 217; in models of planetary motion, 23–26. *See also* iconography of Moon; light of Moon; lunar features; lunar surface, characteristics of; lunar surface, descriptions of; lunar surface, representations of; mapping of Moon; substance of Moon

Moon, ideas of: as another Earth, 14–15, 20–22, 96, 110–120, 140–143, 149, 184–190, 221; as deity, 13–14, 19, 65; of diverse substance, 77, 79; Gilbert's, 100; as glassy, 33, 78; as impure, 16–17, 27, 33, 69; as inhabited, 20–22, 27, 33–34, 139–140, 142–149, 189–190; Leonardo's, 96; as mirror, 17, 33, 95–96; as mythical place, 14, 19, 34–35; as Paradise, 146; as perfect orb, 20; post-Galilean, 128–132; Ptolemaic, 41; as self-luminous, 71, 75; as sphere, 70–71; as uninhabited, 203

Moore, Patrick, *New Atlas of the Universe*, 3–6

mountains of Moon, 35, 117–121, 123

Musaeus, 14

mythology, Greek, 13–15

naming, conventions of, 5–9, 209–218, 223–224; Bayer, 105–106; Galileo, 127–128; Gassendi, 155–156; Gilbert, 98–104; Hevelius, 184–190, 205–206; Plutarch, 34–35; politics of, 98–104; Riccioli, 198–208; Van Langren, 155, 160–168

Naturalis historia (Pliny the Elder), 41–42, 52, 56–58, 60–61

naturalism, 64, 79–82, 85–97, 123–128

Natural Magic (della Porta), 120

Neptune, 210–211

New Atlas of the Universe (Moore), 3–6

Newes from the New World . . . (Jonson), 138–140

Newton, Isaac, 206

New World, exploration of, 101–102, 109–110

New World, Moon as, 98–104, 121–122, 150, 175

Nicolson, Marjorie Hope, 135–136

Nova demonstratio immobilitatis terrae . . . (Grandami), 192–193

observation, 74, 86–95, 175–176, 178, 181. *See also* naturalism

observatory, Hevelius's, 170, 239 n. 2

On Nature (Parmenides), 16–17

On the Light of the Moon (Al-Haytham), 72, 75

On the Nature of the Marks Seen on the Surface of the Moon (Al-Haytham), 72–76

Oresme, Nicholas, *Livre du ciel et du monde,* 78

Orlando furioso (Ariosto), 136–138

Orpheus, 14

Ortelius, Abraham, 127; *Theatrum orbis terrarum,* 105

Panofsky, Erwin, 94

Paradise Lost (Milton), 150

Parmenides, *On Nature,* 16–17

patronage, 81–82, 85–86, 93–94, 159–160

Peiresc, Nicolas-Claude Fabri de, 151–156

About the Author

Scott L. Montgomery is a geologist, author, and independent scholar who has written widely on various topics related to the history of science, science and art, language studies, education, and cultural criticism. His most recent books include *Minds for the Making: The Role of Science in American Education, 1750–1990* (Guilford Press, 1994), *The Scientific Voice* (Guilford Press, 1996), and *Science and Translation: Movements of Knowledge in Culture and Time* (University of Chicago Press, 1999). He currently lives with his wife, Marilyn, and his two sons, Kyle and Cameron, in Seattle, Washington.